T0291993

Corticonics

Corticonics

Neural circuits of the cerebral cortex

M. ABELES

Department of Physiology, School of Medicine
The Hebrew University of Jerusalem

The right of the
University of Cambridge
to print and sell
all manner of books
was granted by
Henry VIII in 1534.
The University has printed
and published continuously
since 1584.

CAMBRIDGE UNIVERSITY PRESS

Cambridge
New York   Port Chester   Melbourne   Sydney

CAMBRIDGE UNIVERSITY PRESS
Cambridge, New York, Melbourne, Madrid, Cape Town, Singapore,
São Paulo, Delhi, Dubai, Tokyo, Mexico City

Cambridge University Press
The Edinburgh Building, Cambridge CB2 8RU, UK

Published in the United States of America by
Cambridge University Press, New York

www.cambridge.org
Information on this title: www.cambridge.org/9780521376174

First published 1991

A catalogue record for this publication is available from the British Library

Library of Congress Cataloguing in Publication Data

Abeles, Moshe, 1936—
Corticonics : neural circuits of the cerebral cortex / M. Abeles.
p.   cm.
Includes bibliographical references.
Includes index.
ISBN 0-521-37476-6 (hardback). – ISBN 0-521-37617-3 (pbk.)
1. Neural circuitry.   2. Neural networks (Computer science)
3. Cerebral cortex – Computer simulation.   I. Title.
[DNLM: 1. Cerebral Cortex – anatomy & histology.   2. Cerebral
Cortex – physiology.   3. Neural Transmission.]
QP363.3.A24   1991
612.8'25 – dc20
DNLM/DLC
for Library of Congress   90-2035
                                                                        CIP

ISBN 978-0-521-37476-7 Hardback
ISBN 978-0-521-37617-4 Paperback

to Gabi

Contents

Preface

Comprehending how the brain functions is probably the greatest intellectual and scientific challenge we face. We have virtually no knowledge of the neural mechanisms of relatively simple brain processes, such as perception, motor planning, and retrieval of memories, and we are completely ignorant regarding more complex processes, such as cognition, thinking, and learning.

The cerebral cortex (and particularly its most recently developed part, the neocortex) is considered essential for carrying out these higher brain functions. Although the neocortex has been divided into many subareas, each subarea contains many small modules composed of the same building elements, which are interconnected in essentially the same manner. I subscribe to the belief that the same basic computations are carried out by each small region of the neocortex. Understanding the nature of these computations and their neurophysiological mechanisms is a major challenge for science today.

There are many strategies for attacking these questions. Some researchers believe that one must first prepare detailed descriptions of all the types of neurons and of their internal and external connections. Others believe that the most urgent need is to construct appropriate models to describe how the brain might work. Between these extreme bottom-up and top-down strategies there is room for a multitude of intermediate approaches. The work reported here exemplifies one such approach. It combines anatomical observations, electrophysiological measurements, and abstract modeling in order to obain a quantitative description of cortical function. Because the emphasis in this work is on the *quantitative,* I use the term "corticonics," in which "cortex" is conjugated in a fashion similar to "electronics." This book is intended for students and colleagues who are interested in quantitative analysis of the structure and function of the cerebral cortex. Its organization derives from a graduate course I taught at the Hebrew University of Jerusalem in 1982 and 1984, and consequently it has taken on the form of a text-book. It does not give a comprehensive review of any of the topics it touches, but rather describes in didactic form all the facts and methodology required to understand the subject matter.

The reader is expected to have only a basic understanding of neurobiology (i.e., familiarity with neurons, action potentials, and synaptic potentials) and basic knowledge of mathematics (the ability to solve simple integrals and make simple calculations of probability).

While writing the text, I considered the possible readership from the fields of neurobiology, computer science, medicine, psychology, and artifical intelligence. Catering to such a wide range of readers necessarily implies that one may find sections dealing with material that one already knows, or sections that are of little interest.

Chapter 1 presents a review of the anatomy and embryogenesis of the cerebral cortex. The reader who is familiar with this topic may skip to the last section, dealing with the quantitative aspects of cortical anatomy.

Chapter 2 shows how to estimate the probability of contact between neurons. It is intended principally for those who are interested in studying cortical structure on a quantitative basis.

Chapters 3 and 4 deal with neurophysiological methods for studying the relations between pairs of neurons and how these relations can be used to assess the function of complex neuronal circuits.

Chapter 5 presents a general review of the analysis of activity in populations of neurons. It deals with random networks, recurrent networks, and feed-forward networks.

Chapter 6 takes us back to the anatomy of the cortex. It deals with the plausibility of connections between populations of cortical neurons.

Chapter 7 deals with my own "pet" model of synfire chains.

Throughout the text I indicate which sections cover material standard to a given field and which are more specialized, so that a reader may determine which sections to skip.

This book deliberately does not deal with the topic of learning. Learning mechanisms are studied intensively by both modelers of neural networks and molecular neurobiologists. However, while preparing this book (1985–8), I believed that not enough was known about the relationship between learning and the cortex to justify the inclusion of a quantitative section on cortical learning.

A textbook on a quantitative subject calls for exercises. Rather than collect them at the end of each chapter, I chose to spread them out in appropriate places throughout the text. These should be regarded by the reader as aids for determining whether or not the argument leading up to the exercise was completely clear. If so, one should be able to solve the exercise at that point. The exercises also indicate the types of real problems that can be solved with the techniques developed in this book. Some colleagues who read the manuscript found that they could not afford the time required to crank the mathematical machinery needed to solve the exercises, but they nonetheless wanted to know the answers. For their sake, I have provided the answers in the Appendix.

Occasionally I encountered relevant questions that could not be solved simply by the methods developed in this book. These questions are reffered to as "problems." They represent small research topics. When I taught this material as a graduate course, my students could choose any of these problems as a take-home examination. I found many of the students' solutions illuminating and have used some of them (with the appropriate acknowledgments) in the text.

Earlier versions of this work were sent to several colleagues, whose feedback was valuable for preparing the version presented here. I would like to thank all of them, and in particular A. Aertsen, F. de Ribaupiere, E. E. Fetz, G. L. Gerstein, M. H. Goldstein, Jr., H. Gutfreund, M. Gutnick, and H. Sompolinsky. Special thanks are due to I. Nelken, whose comments and suggestions helped me improve many sections throughout the text. In addition, I would like to thank D. Diengott and S. Kamovitz for their help in language editing, V. Sharkansky for her help with the drawings, and H. Reis for patiently typing and retyping the many versions and alterations of this text. I am particularly thankful to A. Shadmi for taking on much of the technical and administrative load associated with getting this work finished.

Many institutions and agencies have contributed in different ways toward the completion of this work. The idea of putting bits and pieces of knowledge from anatomy, neurophysiology, and mathematics together into a new topic, "corticonics," arose while I spent a year at the Institute de Physiologie in Lausanne in 1980. I thank both the institute and Hoffman–La Roche for their hospitality and support during that year. The bulk of the text presented here was written while spending a most fruitful year at the University of Pennsylvania in 1986. I thank Professor G. L. Gerstein and the System Development Foundation for giving me that opportunity.

Although I felt that this book would not be complete without a section on recent developments in neural network modeling, I did not feel knowledgeable enough on these subjects to write even a superficial text. In 1988 I spent a year at the Institute for Advanced Studies in the Hebrew University of Jerusalem as a member of a study group on spin-glass theory and its applications in physics, biology, and mathematics. I am most grateful to all the other members and guests of this group for their help in introducing me to this topic. Some of my own research that is described in this book was supported by the USA–Israel Binational Science Foundation and by the Fund for Basic Research administered by the Israel Academy of Sciences and Humanities.

A major source of encouragement and support leading to this work has been The Israel Institute for Psychobiology – The C. E. Smith Family Foundation, without which this work could not have been completed.

1
Anatomy of the cerebral cortex

This chapter presents a brief overview of the macroscopic structure, the microscopic structure, and the embryological genesis of the cerebral cortex. In the last section of this chapter, the quantitative aspects of the cortical morphology are considered. It should be noted that a comprehensive survey of all the known facts about the cortex would be an impossible task. Any selection of facts necessarily reflects the personal opinions of the author regarding the relevance of particular facts and issues. The selection of topics and facts presented here is certainly biased by the general aims of the later chapters and by my view that the brain is essentially deterministic on a large scale, but probabilistic on a small scale.

In studying this chapter, the reader is expected to understand the "classical" views of the cortical anatomy and microscopy, to understand methods and concepts that are dealt with in current neuromorphological research, and to be able to interpret microscopic images in a manner suitable for calculating the probabilities of contact, as described in Chapter 2. Most of this chapter is intended for the reader with little background in neurobiology. The reader who is familiar with the cortical structure may skim this chapter, paying detailed attention only to the tables in Section 1.5.

1.1 Macroscopic structure

In lower forms of vertebrates, fish, amphibia, and reptiles, the anterior part of the brain (the forebrain) is relatively undeveloped. In these animals the forebrain appears to be related to the olfactory system. It is thus terms the "rhinencephalon" (smell-brain). Only in mammals has the forebrain developed to the extent that it comprises most of the mass of the central nervous system. Those parts of the forebrain that seem to be homologous with the forebrain in the lower vertebrates retain the

1

name rhinencephalon. One must remember that in mammals the rhinencephalon has developed along with the evolution of the forebrain and that it has acquired many additional connections with other parts of the forebrain. In mammals, it is customary to term those parts of the cortex that can be recognized in lower vertebrates the "archipallium" (the old mantle), and the parts that have developed mostly in mammals are called "neopallium," or neocortex.

As we climb up the mammalian phylogenetic tree, we find that the part of the brain that has developed most extensively is the forebrain, and within it the neopallium shows the highest rate of evolution. Because the cortex grew much faster than the skull, it acquired many folds. Cortical folding is not merely a mechanical necessity; it has a significant functional advantage as well. Underneath the cortex is a great mass of fibers (the white matter). Over 99 percent of these fibers connect one cortical region to another. These fibers would have been much longer had the cortex not been folded. Figure 1.1.1 shows a frontal cut through a human brain. The ratio between the cortical mantle and the underlying white matter can be seen in this figure. One can also appreciate how many more fibers would be needed if the cortical folds were stretched out and we would still want to connect all parts of the cortex to one another.

In the mammalian brain the very deep folds of the cortex are called fissures, and the shallower folds are called sulci. The folded cortex between two adjacent sulci is called a gyrus. These gyri usually take the form of elongated convolutions. For the sake of completeness, Figure 1.1.2 shows different views of the human cerebral cortex.

When comparing cortices of different individuals, one can easily recognize the deep fissures and some of the deeper sulci. The positioning of the shallower sulci can vary considerably among brains of individuals from the same species. Figure 1.1.2 represents an "idealized" average figure of the exterior of the human forebrain.

The cortex in lower mammals is almost completely flat, generally becoming more and more convoluted as we climb up the phylum. The cortices of mice and rats have practically no folds, whereas monkeys have homologues for most of the major folds that humans have. This general rule has several exceptions, however. The cortex of the owl monkey, for instance, is almost flat (to the convenience of electrophysiologist, who like to map its functions), whereas dolphins and whales have cortices as convoluted as those of humans.

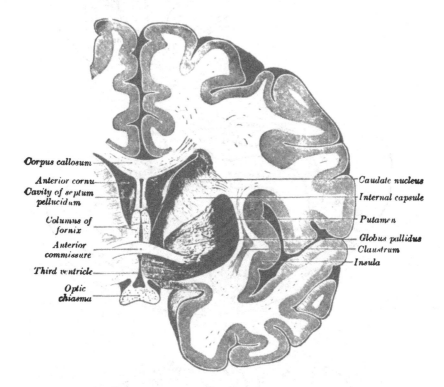

Corpus callosum

Anterior cornu

Cavity of septum
pellucidum

Columns of
fornix

Anterior
commissure

Third ventricle

Optic
chiasma

Caudate nucleus

Internal capsule

Putamen

Globus pallidus

Claustrum

Insula

Figure 1.1.1. Frontal section through the human forebrain. [From H. Gray, *Anatomy of the Human Body,* 26th ed., 1954, with permission of Lea & Febiger.]

Despite the enormous changes in the amount of cortex that have occurred throughout phylogeny, the internal structure of the cortex has changed little. Figure 1.1.3 illustrates pieces of cortex taken from different animals. Alongside each piece is given the ratio between the brain size of that animal and the brain of a mouse. Though the human brain is 3,400 times larger than that of the mouse, its cortex is only three times thicker. A gross examination of a cortical slice shows that it is composed of several layers. The different appearances of the layers are due to the varying proportions of cells and fibers. The different layers also tend to contain neurons of different sizes and types. The same pattern of lamination is seen in different neocortical areas of a given species, as well as in neocortices of different mammals. This suggests that all pieces of cortex perform basically the same types of transactions on the information they

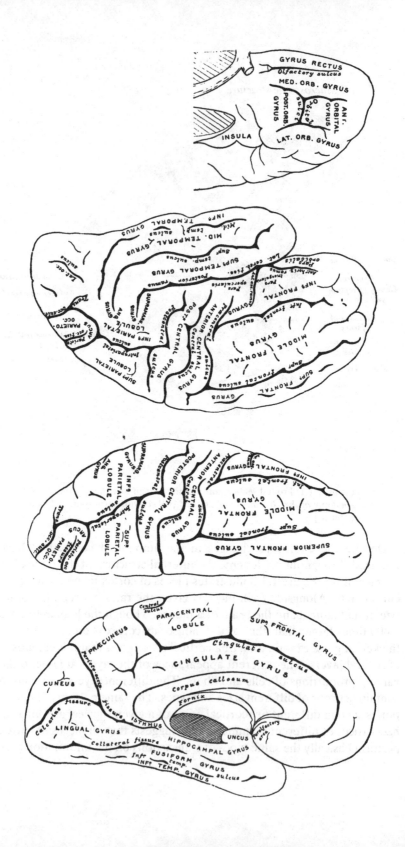

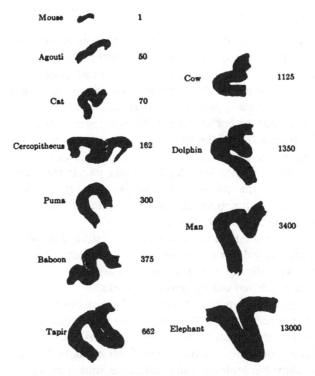

Figure 1.1.3. Comparing cortices of different mammals. Sections through pieces of cerebral cortices of various mammals drawn on 1:1 scale. Although cortical thickness varies only slightly, brain size (figure to the right of each cortical piece) varies considerably. Brain size is given as a ratio between a brain's weight and the weight of the mouse brain. [From C. U. A. Kappers, G. C. Huber, and E. C. Crosby, *The Comparative Anatomy of the Nervous System of Vertebrates, Including Man*, 1936, with permission of Macmillan & Co., Ltd.]

receive and that the differences among brains lie mainly in the numbers of transactions that can be carried out concomitantly. The idea of invariant cortical transaction is further supported when the microscopic structures of the cortices of various mammals are compared, as will be seen in later sections.

Figure 1.1.2. The sulci and gyri of the human left hemisphere. The front end is pointing to the right. From bottom to top: view from the right; view from above (the hemisphere is drawn upside down); lateral view; view from below (or the frontal lobe only). [Based on H. Gray, *Anatomy of the Human Body*, 26th ed., 1954, with permission of Lea & Febiger.]

1.2 Types of neurons

Cortical tissue is composed mainly of two types of cells: the nerve cells (or neurons) and the neuroglial cells (or glia). Although the glia outnumber the neurons by far, their function is not fully understood. They are thought to have some metabolic supportive role. Some glial cells (oligodendrocytes) provide the myelin sheath that wraps many of the axons in the white matter and the cortex. Others (astrocytes and microglia) control the ionic composition of the interstitial space of the brain. Some glia may exchange RNA and proteins with neurons [Hyden, 1967]. The available evidence indicates that glia do no take part in the interactions between neurons on the millisecond scale, yet may play a role in slow modulations of neural function. Glia play an important role in directing the development and growth of the brain.

It is believed that the intricate processing and storage of information in the brain are carried out by neurons. We shall thus deal exclusively with neurons, with the connections among neurons, and with the processes that can be carried out by neural networks.

It was not until the nineteenth century that scientists realized that tissues are composed of cells. Previously, the soft nature of the brain tissue had not lent this organ to detailed histological study. Only toward the end of the nineteenth century were fixation methods sufficiently developed to allow hardening of the brain tissue, thus enabling researchers to prepare thin sections that could be studied carefully under the microscope. Scientists of that period debated whether each neuron was a separate entity or whether all the neurons were connected to form one enormous syncytium. It was the greatest brain histologist of all time, Ramon y Cajal, who persuaded most of the scientific community that each neuron is a closed unit and that these units interact with each other at specialized contacts called synapses. See Van der Loos [1967] for a thorough description of this controversy.

This issue was not ultimately resolved until recent years, when synapses were examined under the electron microscope. It was observed that two membranes exist in the synapse, one belonging to the presynaptic neuron, and one to the postsynaptic neuron. Between these two membranes there is an interstitial space: the synaptic gap. Ironically, electron-microscopic (EM) studies in recent years have revealed several cases in which adjacent neurons have been connected by gap junctions through which materials can flow freely from one neuron to another. In the mammalian cortex, such gap junctions are found only rarely, and they will not be treated further here.

Cortical neurons may be divided into several categories. The classical categorization of neurons was based mainly on the morphology of the cell body and its dendritic tree, and to a lesser extent on the morphology of the axon and its branches. In recent years, several other criteria have been proposed for categorizing neurons. Among these are the brain region to which the axon projects, the type of synapses made by the neuron, the transmitters released by the neuron, and several other biochemical criteria (the presence of certain neuropeptides, or enzymes, or immunological properties). However, the classical categorization is most commonly employed. It was, and still is, based on the morphology of the cell as seen after Golgi impregnation.

A detailed description of Golgi impregnation is given in Section 1.3.1. At present it is sufficient to note that it is based on a discovery made by Golgi at the end of the nineteenth century: When fresh brain tissue is placed in a concentrated solution of silver salts, these tend to precipitate in some neurons. When they do so, the precipitate fills up the soma (cell body) and many of the branches that belong to the cell. Only small percentages of the cells are impregnated, and one can therefore obtain a clear view of what seems to be the entire extent of individual neurons.

The degree of detail revealed determines the number of distinct categories that one can see. Thus, Lorente de no [1933] described more than sixty cell types, but Sholl [1956] described only two types: pyramidal and stellate cells. The pyramidal cell has a cell body shaped like a pyramid, with local dendrites extending from its base, and one long apical dendrite extending from the pyramid's apex toward the cortical surface, where it branches out extensively. According to Sholl, all nonpyramidal neurons should be considered stellate cells. The typical stellate cell's body has the form of a star (hence its name), with dendrites extending from all aspects of the soma and branching out extensively around the cell body.

There seem to be two approaches to the study of the structure of the cortex: a maximalist approach in which investigators believe that they must create as many categories as possible, and a minimalist approach in which investigators try to limit the number of categories as much as possible. In this work I adopt the view that for every problem there is an adequate classification scheme. For instance, if one is interested only in assessing the cortical domains from which a neuron derives its input, the two categories of Sholl are adequate. In his scheme, the pyramidal neuron differs from the stellate neuron in that the stellate cell derives its excitation only from axons that reach the vicinity of the cell body, whereas the pyramidal cell is affected by axons within its somatic vicinity and by axons at the superficial layers of the cortex that may excite its apical dendritic tree. On the other hand, if we are interested in the local interactions occurring in the cortex, the shape of the dendritic tree alone is insufficient, and we must also consider the morphology of the axon through which the neuron delivers its effects to other cells.

Golgi pointed out that for every nervous structure there are two fundamentally different cell types: one whose axon extends from the structure to deliver its effects to other regions (Golgi type I cells), and one whose axon remains within the structure to exert its entire effect locally (Golgi type II cells). However, the cortex shows an extraordinary feature. The axons of all the projecting neurons also have an extensive collateral tree that affects the local neurons.

Lorente de No [1949] stressed the importance of the distribution of axons in the cortex. In this respect he recognized four cell types in the cortex:

1.   Neurons with axons that grow downward and usually depart the cortex into the white matter
2.   Neurons with axons that branch out mainly in the vicinity of the cell body
3.   Neurons with axons that grow toward the surface of the cortex
4.   Neurons with horizontal axons

This classification is not unrelated to the classification by dendritic-tree morphology described earlier. Almost every pyramidal cell has an axon extending into the white matter; according to Braitenberg [1977], they all do. Only rarely will a stellate cell have an axon that extends into the white matter.

The cells having axons that grow toward the cortical surface were first described by Martinotti. They are concentrated mostly in the deeper cortical layers. The cells with horizontal axons are found almost exclusively in the most superficial layer. They are densely packed during early stages of embryogenesis, but become widely separated as the cortex develops.

According to Braitenberg [1977], there are three major types of neurons:

1. The pyramidal cell, which is characterized by an apical dendrite and an axon that always extends from the cortex into the white matter. Each axon has, in addition, many collaterals that have many synapses on neighboring neurons.
2. The stellate cell, whose dendritic and axonal branches extend only locally.
3. The Martinotti cell, situated at the depth of the cortex, whose axon grows toward the cortical surface, extending few collaterals on its way up, but branching out profusely under the cortical surface, where it makes synapses on the apical dendritic tufts of the pyramidal cells.

Many investigators do not consider the horizontal cells significant for cortical function because of their low density in the adult brain.

If we wish to consider the physiological effects of the cortical cells on each other, we may wish to consider two main categories: excitatory and inhibitory cells. Excitatory cells release transmitters at their synaptic ends that, on contact with the postsynaptic membrane, create currents that depolarize the postsynaptic cell. Inhibitory cells release transmiters that tend to hyperpolarize the postsynaptic cell or to increase the conductance of the postsynaptic membrane to chloride, thereby diminishing the effects of the depolarizing currents generated by the excitatory synapses. In theory, one could imagine a situation in which a certain transmitter would have excitatory effects on one type of cell and inhibitory effects on the other. However, such dual effects are not known to exist in the normal cortex. It is also believed that in the cortex, all the synapses that a given neuron makes release the same neurotransmitters (Dale's principle). There are several notable exception to these schematic rules. In tissue cultures (and in vivo), a given transmitter substance may have opposing effects on different neurons. This is not known to occur in the regular synaptic transmission in the cortex. Some nerve endings release more than one substance. Usually one of the substances is a small molecule that may act as a classical transmitter, and the others are macromolecules that are believed to act in a more diffuse manner (in time and space) and often are referred to as neuromodulators.

Today, most investigators agree that for the analysis of cortical function, it is safe to assume that excitatory and inhibitory cells can be categorized according to the types of transmitters they release at their synaptic endings.

EM studies of the synapses in the cortex have revealed the existence of two major types of synapses. The exact features that distinguish the two types will depend on the type of fixation used while preparing the tissue for the EM study. When aldehydes are used in the fixation process, one type of synapse contains a homogeneous population of small rounded vesicles, whereas the other type contains a mixed population of round and flattened vesicles. When fixed with osmium and stained with phosphotungstic acid, one type of synapse shows marked differences in thickness (and fine structure) between the presynaptic and postsynaptic membranes, whereas in the other type the differences between the two membranes are small. These two types of synapses are known as Gray type I and Gray type II synapses [Gray, 1959], or synapses with round vesicles and synapses with pleomorphic vesicles, or asymmetric and symmetric synapses. For the cerebellum, where the same types of synapses exist, there is direct electrophysiological evidence that the asymmetric synapses are excitatory, and the symmetric synapses are inhibitory. In the cortex there are several indirect indications that the same rule holds. See Colonnier [1981] for a review of this issue. If we accept this rule, it would be logical to categorize cortical neurons according to whether they make asymmetric or symmetric synapses. In the following discussion, symmetric synapses are assumed to be inhibitory, and asymmetric synapses are assumed to be excitatory.

Fortunately, the classification according to synaptic shape (and function) is strongly correlated with the other morphological categorizations. The axon of the pyramidal cell is always involved with asymmetric (excitatory) synapses. For most of the stellate cells, the axon is involved with symmetric synapses. The stellate cell whose axons makes asymmetric synapses appears to have dendritic morphology that differs from that for the other (inhibitory) stellate cells. The difference lies in the densities of dendritic spines that are attached to the dendrites of the stellate cells. A dendritic spine (Figure 1.2.1) is a mushroomlike appendage that is attached to the dendritic shaft by a narrow stalk. The dendrites of the stellate cell whose axon generates asymmetric (excitatory) synapses have many spines attached to them. Most of the synapses that other cells generate on these dendrites are found on the spines. These cells are called spiny stellate cells. They are presumably excitatory. In contrast, the inhibitory stellate cells have very few spines on their dendrites, or none at all. They are called smooth stellate cells, or sparsely spinous stellate cells. The term "smooth" does not mean that the dendrite has a regular, cylindrical form; on the contrary, the smooth dendrites have many varicosities, and their appearances are highly irregular. It is appropriate to note here that the pyramidal cells whose axon generates asym-

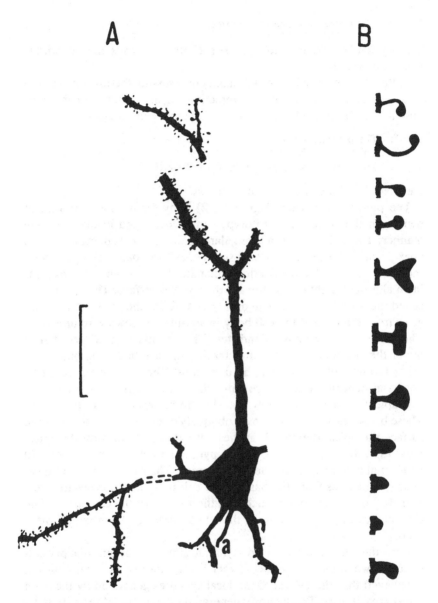

Figure 1.2.1. Dendritic spines. A: A small pyramidal cell from upper layer II–III. [From M. L. Feldman, "Morphology of the Neocortical Pyramidal Neuron," in A. Peters and E. G. Jones (eds.), *Cerebral Cortex, Vol. 1*, 1984, with permission of Plenum Publishing Corp.] B: Different spine shapes. [From A. Peters and I. R. Kaiserman-Abramof, "The Small Pyramidal Neuron of the Rat Cerebral Cortex. The Perikaryon, Dendrites, and Spines," *American Journal of Anatomy*, 127:321–56, 1970, with permission of Alan R. Liss, Inc.]

metric synapses and are known to be excitatory cells also have dendrites with many spines.

In this book we are interested mainly in assessing the functions of the local excitatory and inhibitory interactions in the cortex. For this purpose it is useful to deal with the following main cell categories:

1. The pyramidal cell
2. The spiny stellate cell
3. The smooth (or sparsely spinous) stellate cell

Detailed descriptions of these cells follow.

The pyramidal neuron (Figure 1.2.2) is by far the most prominent neuron in the cortex. Its prototype can be described in the following manner: The cell body has a triangular profile (hence its name), with the apex pointing toward the cortical surface, and the base pointing toward the white matter. A thick dendrite emanates from the apex and extends in an almost straight course toward the cortical surface. This dendrite is called the apical dendrite. The other dendrites of the pyramidal neuron grow from the base of the cell body in an oblique, downward direction. These are called the basal dendrites. Their shafts form a sort of skirt under the cell body. Shortly after leaving the cell body, the basal dendrites branch out profusely in all directions. The volume embraced by the basal dendritic tree encapsulates the cell body from all directions. The apical dendrite grows few branches on its way toward the surface. These branches are oriented in an obliquely outward direction. Near the surface the apical dendrite branches out extensively to form the apical dendritic tuft. The dendrites of the pyramidal cell are thickly covered with small appendages, the dendritic spines. The axon of the pyramidal neuron emanates from the base of the cell body in a downward direction. It grows numerous branches in the vicinity of the cell body, while the main axon proceeds toward the white matter and leaves the cortex there.

In all the known cases, the postsynaptic effects of the axons of pyramidal cells (when they reach their target) are excitatory. For that reason it is believed that the effects of the local synapses generated by the axon are also excitatory. The synapses generated by the axon of the pyramidal cell are of the asymmetric type.

The distribution of synapses received by the pyramidal cells is also typical. The cell body and the initial shafts of the dendrites receive almost exclusively symmetric synapses. One tends to find strings of symmetric synapses on the initial segment of the pyramidal cell axon as well.

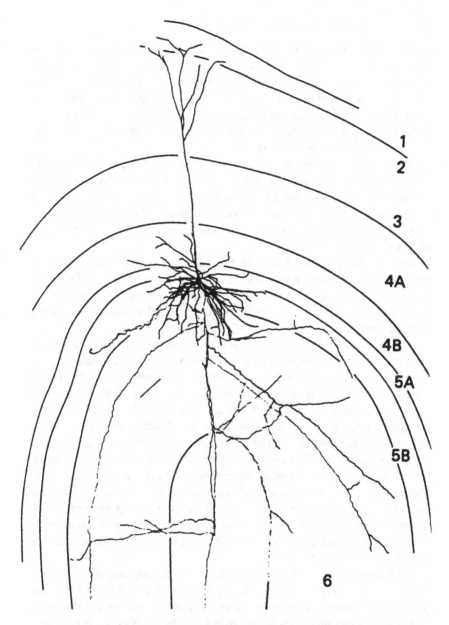

1
2
3
4A
4B
5A
5B
6

Figure 1.2.2. A pyramidal neuron from the visual cortex of the cat. The total extent of the cell was demonstrated by injecting a dye (HRP) into the cell. [From K. A. Martin and D. Whitteridge, "Form, Function and Intracortical Projection of Spiny Neurons in the Striate Visual Cortex of the Cat," *Journal of Physiology* (*London*), 353:463–504, 1984, with permission of The Physiological Society.]

Synapses that are formed on remote dendrites are found primarily on dendritic spines. These synapses are mostly of the asymmetric type. Synapses that are made on remote dendritic shafts are of both types, with slightly more of the symmetric type. This arrangement suggests that the major role of the inhibitory effects on the pyramidal cell is to suppress its activity completely and that excitation is achieved only by the summation of many synaptic inputs.

One can find numerous variations of this prototypical, "classical" description. In deep cortical layers one can find cells having all the foregoing properties, and yet the cell body will be elongated and in many cases rotated sidewise to various degrees. The "apical dendrite" may start in a sidewise direction, but shortly afterward turn in the usual outward direction.

Some of the pyramidal cells in the deep layers may have an apical dendrite that does not quite make it to the surface. In the middle of the cortex, one finds "pyramidal cells" in which the basal dendrites extend from the entire cell body, not just from the base. These have been called "star pyramids."

Some pyramidal neurons will have an axon that, in addition to the local branches in the vicinity of the cell body and the main branch that leaves the cortex, will also have a major intracortical collateral that branches out profusely in a volume that does not contain its cell body or any of its basal dendrites. Such axonal branches, which remain within the gray matter, may project sidewise to branch off in areas that can be as far as 2 mm from the cell body (Figure 1.2.2). In other cases these additional intracortical branches may branch off in cortical layers above (or below) the layers in which the cell body and its basal dendrites lie. Usually the axons of the pyramidal cell emanate from the base of the cell body, but in some cases the axon may grow from a main shaft of one of the basal dendrites rather than directly from the cell body. A full description of the morphology of the pyramidal neuron can be found in a recent review by Feldman [1984].

The spiny stellate cells (Figure 1.2.3) have multipolar somata, with dendrites extending from the poles in all directions. The dendrites are covered with spines. The axon leaves the cell in a downward direction and branches out profusely in the vicinity of the cell's body. It may also grow a collateral that emits many branches in more superficial cortical layers. The synapses generated by the axon are asymmetric (presumably excitatory). The spiny stellate cell, like the pyramidal cell, receives on its cell body and its initial dendritic shafts almost exclusively symmetric

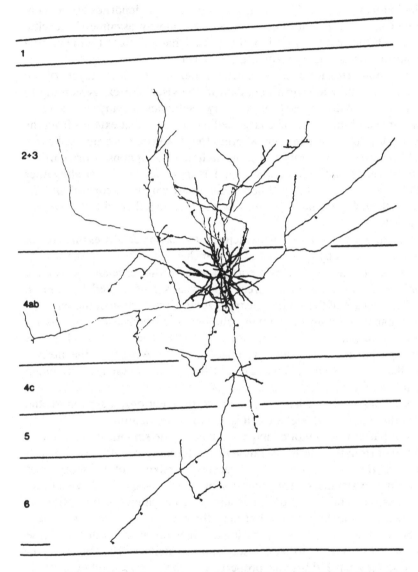

Figure 1.2.3. Spiny stellate cell. The full extent of this cell, from the visual cortex of the cat, was demonstrated by intracellular injection of a dye (HRP). [From C. D. Gilbert and T. M. Wiesel, "Laminar Specialization and Intracortical Connections in Cat Primary Visual Cortex," in F. O. Schmitt et al. (eds.), *The Organization of the Cerebral Cortex*, 1981, with permission of MIT Press.]

(inhibitory) synapses. The synapses on the remote dendrites are concentrated mostly on dendritic spines and are mostly asymmetric (excitatory). Of the small proportion of synapses that are made on the remote dendritic shafts, the majority are symmetric.

The spiny stellate cells are concentrated in the middle layers of the cortex, and they are most abundant in the visual cortex. According to Gilbert and Wiesel [1981], there are two subtypes of spiny stellate cells: the large and the small. The large cell has an axon that extends from the cortex into the white matter, whereas the small cell has only local cortical branches. The number of spiny stellate cells in nonsensory cortices can be very small, and in some animals they may be absent altogether [Peters and Kara, 1985b]. When dealing only with local cortical interactions, the spiny stellate cells may be grouped together with the pyramidal cells.

The smooth stellate cell is characterized as having dendrites that are not densely covered by spines. Its axon branches out only within the cortex, although it often embraces a cortical volume that exceeds the volume embraced by the dendritic tree. The synapses generated by smooth stellate cell are of the symmetric type. A large proportion of the symmetric synapses contain the enzyme glutamic decarboxylase, which is required to produce γ-aminobutyric acid (GABA) [Freund et al., 1983]. A large proportion of the smooth stellates absorb GABA from the interstitial fluid [e.g., Homos, Davis, and Sterling, 1983]. These facts are taken to mean that these cells utilize GABA as their transmitter, and because GABA has an inhibitory effect on the cortical neurons, that supports the idea that the smooth stellate cell is an inhibitory neuron.

The pattern of synaptic inputs received by the smooth stellate cells is different from that for the pyramidal and the spiny stellate cells. The cell body of the smooth stellate cell receives a mixture of symmetric and asymmetric synapses. The majority of these synapses are symmetric. The remote dendrites also receive a mixture of symmetric and asymmetric synapses on their shafts, but here the asymmetric synapses prevail.

Some stellate cells have occasional spines on their dendrites. These are usually referred to as sparsely spinous stellate cells. In other respects, they have the same properties as the smooth stellate cells. It seems that the sparsely spinous stellate cells tend to shed their spines throughout life, and so it is not clear that these form a distinctive class.

The smooth stellate cells show great variety in their dendritic and axonal morphology, and they have been divided into many subgroups by several scientists, each having somewhat different criteria for the classifi-

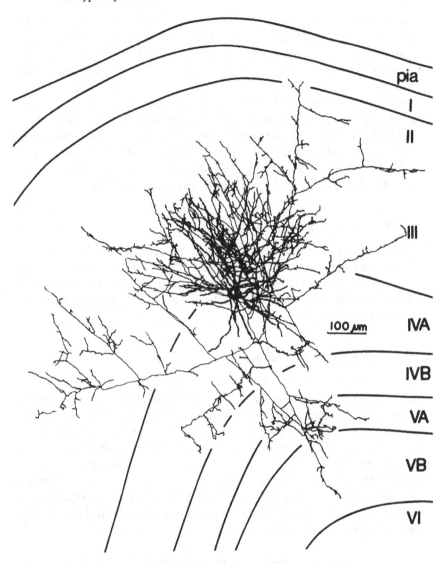

Figure 1.2.4. A smooth stellate cell. The full extent of this cell, from the visual cortex of the cat, was demonstrated by intracellular injection of a dye (HRP). [From P. Somogyi, Z. F. Kisvarday, K. A. C. Martin, and D. Whitteridge, "Synaptic Connections of Morphologically Identified and Physiologically Characterized Large Basket Cells in the Striate Cortex of the Cat," *Neuroscience*, 10:261–94, 1983, with permission of Pergamon Press PLC.]

cation. Not all the smooth stellate cells are really star-shaped. Some have only two dendrites (bipolar cells), or even only one dendrite (monopolar cells). In some cells the dendrites form two groups, one pointing outward and the other pointing inward (bitufted cells). The details of the axonal shape and its distribution have also been used for classification. Many of the smooth stellate cells tend to give synapses preferentially on the pyramidal cell body. These have been called basket cells (Figure 1.2.4). The so-called chandelier cell has an axon that forms synapses exclusively with the initial segments of the axons of pyramidal cells [Somogyi, 1977]. The interested reader can find comprehensive reviews of these cells in the work of Peters and Jones [1984].

1.3 Microscopic organization

This section describes the ways in which the various cell types are organized in the cortical layers, the division of the cortex into zones, and the connections between the cortex and other structures. It starts with a brief description of the staining methods used for studying the morphology of the cortex.

1.3.1 Staining methods

Cells cannot be observed directly under the microscope because they are translucent. Although in recent years phase-contrast microscopy has advanced to the stage where the boundaries of living cells can be seen, this method is of little use for the brain, where cell bodies and processes are packed together tightly. Therefore, study of the microscopic structure and organization of the cortex must rely heavily on observing cells after staining. The various staining techniques highlight different aspects of the cell morphology; it is therefore important to understand the properties and limitations of the various techniques. The most frequently used techniques are the following.

1. Nissl stains. Nissl staining is a method that uses dyes that are adsorbed to acidic macromolecules. The nucleic acids are the most strongly stained cellular components. The Nissl stains may color the nucleolus very strongly; next in rank is the nucleus, and the ribosomes of the cytoplasm are only slightly stained. These stains can demonstrate all the cells in the brain, including the cells that make up the blood vessels, the glia, and the neurons. Usually the nucleoli of nerve cells are stained darkly, whereas those of glia are not. This is the most important criterion for distinguishing between these two cell types. The Nissl stains cannot show the axon or the dendrites of the neurons.

2. Golgi impregnation techniques. As mentioned in Section 1.2, Golgi techniques are based on the precipitation of silver chromate salts in the cells. These techniques highlight only a small percentage of the neurons. In a quantitative study of this issue, Pasternak and Woolsey [1975] found that only 1.3 percent of the neurons in rat cortex were impregnated. The Golgi techniques are based on the initial formation of salt crystals in some cells. While the tissue is incubated with the staining solution (for a period of a few days up to several weeks), these initial seeds grow continuously. Because the salt crystal cannot grow through the lipid cell membrane, it is confined to the interior of the cell in which the seed was formed. It is not known why some cells are stained by this technique but many others are not.

The crucial question is whether or not all cell types have equal chances of picking up the stain. Pasternak and Woolsey [1975] compared the distribution of cell sizes recognizable using Nissl stains and the distribution of sizes of cell bodies of neurons seen with the Golgi technique. They found the distributions similar in appearance. However, other investigators have claimed that different variants of the Golgi technique tend to impregnate preferentially different cell types [Meyer, 1983].

Over the years, many variations of the original Golgi impregnation technique have been developed in an effort to improve the impregnation of fine branches (including spines). Although the aim of the Golgi techniques is to render the entire cell visible, that goal is not totally achieved. Usually the myelinated part of the axon is not fully impregnated. In addition, if for some reason the salt crystal ceases to grow at some point in a dendrite, the rest of that dendrite and all its branches will not be impregnated. This issue will be discussed later in regard to the intracellular injection techniques.

3. Weigert stain for myelin. This stain is specific for the fatty materials that make up the myelin sheath. At the white matter, the myelinated axons are packed very densely, and so the tissue appears as a solid black mass. Most of the processes inside the cortex are not myelinated, and therefore it is possible to observe bundles of myelinated axons as well as isolated myelinated branches.

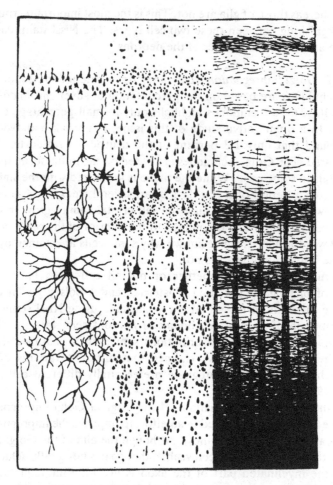

Figure 1.3.1. Cortical slices stained by the Golgi, Nissl, and Weigert methods. [From H. Gray, *Anatomy of the Human Body*, 26th ed., 1954, with permission of Lea & Febiger. After Brodman: from *Luciani's Physiology*, Macmillan & Co., Ltd.]

A schematic representation of results obtained with the Golgi, Nissl, and Weigert methods is shown in Figure 1.3.1.

    4. Reduced-silver methods. These staining methods are selective for neurofilaments that are contained in all the neuronal elements (cell body, dendrites, and axons). The brain is densely packed with neural

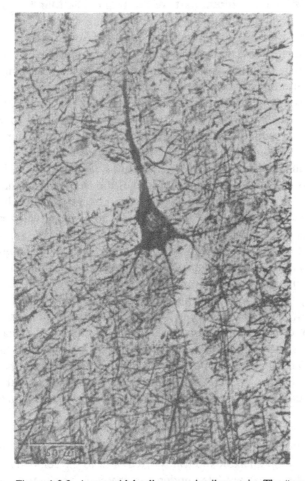

Figure 1.3.2. A pyramidal cell as seen in silver stain. The "empty" spaces are due to blood vessels. [From] D. A. Sholl, *The Organization of the Cerebral Cortex,* 1956, with permission of Methuen, a division of Routledge, Chapman and Hall, Ltd.]

processes, all of which are stained by this technique; therefore, the cortical structure can be resolved only in thin sections. Figure 1.3.2 shows a thin cortical section stained by the silver method as seen in high magnification.

5. Staining of degenerating elements. Several methods of this kind have been used extensively in the past to investigate the major connections between different parts of the brain. The Marchi method, for instance, stains preferentially the decomposing products of myelin. After a given brain region is damaged, all the axons of cells in that region degenerate, and then the myelin sheath of these cells starts to decompose. The Marchi method stains the droplets of decomposing myelin. Because one obtains only patches of stained materials along the course of a single degenerating axon, the method cannot be used to follow the course of a single axon. It has been used extensively to follow tracts of degenerating axons. Most of our knowledge of the major tracts that connect different brain regions derives from such studies.

When the major axonal branch of a neuron is damaged, the cell body usually shows changes in the fine structure of the Nissl-stained material in its cytoplasm. These so-called degenerative changes in the cell body have been used extensively to discover the sources from which major afferent connections with a given target originate. Today such connectivity studies are conducted through the much more sensitive techniques of anterograde and retrograde transport, as described later, but studies of degenerating neural elements can still be of advantage. For instance, White [1978] made a lesion in the somatosensory relay nucleus of the thalamus and, after an appropriate delay, studied the somatosensory cortex under the electron microscope. The terminals of the thalamic afferent axons show typical degenerative changes that are identifiable under the electron microscope. By this method he showed that the thalamic efferents to the cortex establish synaptic contacts with dendrites of all cell types that happen to pass through the termination zones of the axons.

6. Retrograde staining. Certain materials are taken up by the terminal endings of an axon and transported back to the cell body. If the material can be visualized later under a microscope or electron microscope, one can stain all the neurons whose axons terminate in a given zone. The material that is most often employed for this purpose is the enzyme horseradish peroxidase (HRP). The enzymatic properties of this material can be used to convert transparent soluble dyes into an opaque precipitate that can be easily seen. The dye's development is an enzymatic process, and therefore extremely low concentrations of the HRP can be detected [e.g., Mesulam, 1978].

Fluorescent materials are also used for staining neurons by retrograde transport. Although the sensitivity of these dyes is much lower than that of HRP, they have an advantage when one wishes to simultaneously mark several regions and tell which neurons send their axons to which regions [Kuypers et al., 1980].

7. Anterograde staining. Some materials are taken up by the cell body and then transmitted down the axon all the way to its terminals. Thus, one can inject material into a region, wait until it is transported to the terminals, and then discover the locations of these terminals. In the past, the materials most frequently employed for anterograde staining were radioactive amino acids. These are actively transported into the neuron and incorporated into proteins, some of which are transported down the axon. The radioactive material can be visualized by "autoradiography": The brain slice is covered with photographic emulsion; when the radioactive material decomposes, it leaves tiny silver grains in the emulsion. Staining through autoradiography suffers generally from the same shortcomings as the staining of degenerating elements: The stain is patchy, and therefore individual axons cannot be traced. One can visualize with these methods the major tracts leading from the injected site, as well as the locations of the major termination zones of these axons.

In recent years it has been found that neurons can take macromolecules and transport them anterogradely. One can then attach HRP to these molecules using immunohistochemical methods and use its enzymatic properties to demonstrate the entire extent of every projecting axon [Gerfen and Sawchenko, 1984].

8. Intracellular dye injection. In recent years, several investigators have succeeded in impaling cortical neurons with a micropipette containing a dye that is then injected into the impaled cells. When using this most promising technique, it is possible to study the physiological properties of a neuron, inject the dye into it, and then identify the cell later in brain slices. See Kelly and Van Essen [1974], Gilbert and Wiesel [1979], and Martin and Whitteridge [1984] for examples of results obtained with this approach.

When HRP is injected presumably it is transported to all the dendritic and axonal branches of the cell. Because no other cells are stained, it is possible to observe the stained cell to its fullest extent. Pictures of cortical cells obtained in this way seem to show axonal and dendritic trees that are much more elaborate than what had previously been seen with the Golgi technique. Most of the available quantitative information on the morphology of the cortical neurons has been derived with the Golgi technique. It is not possible to stain a single cell with both HRP and

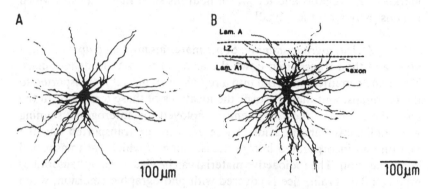

Figure 1.3.3. Comparing Golgi impregnation and intracellular injection of HRP. A: Golgi-impregnated neuron from the LGN of the cat. B: The same type of neuron stained by intracellular injection of HRP. [Based on M. J. Freidlander, C. S. Lin, C. R. Stanford, and S. M. Sherman, "Morphology of Functionally Identified Neurons in Lateral Geniculate Nucleus of the Cat," *Journal of Neurophysiology*, 46:80–129, 1981, with permission of The American Physiological Society.]

Golgi techniques; therefore, it is difficult to give an exact figure for the amount of branching that is missed by the Golgi impregnation technique. Comparison of different cells is difficult in the cortex because of the great morphological variability of cortical neurons. Figure 1.3.3 attempts to compare two neurons of the same type from the lateral geniculate nucleus (LGN) of the cat. One of them was stained by intracellular injection of HRP, and the other by Golgi impregnation. The discrepancy between the two methods can be attributed to two factors: The Golgi method does not impregnate fully myelinated axons, and Golgi impregnation may stop in the middle of a dendrite. Brown and Fyffe [1981] compared some morphometric parameters of spinal cord motoneurons that they injected with HRP with the parameters that had been published by other investigators using other staining techniques. The Golgi-impregnated motoneurons had, on the average, seven main dendrites (ranging from two to fourteen), and the maximal distance of a dendrite from the cell body was 1 mm. The HRP-injected motoneurons had, on the average, 11.6 dendritic trunks (ranging from 7 to 18) that reached up to 1.6 mm from the cell body. Motoneurons injected intracellularly with Procion Yellow or with tritiated glycin showed a number of dendritic trunks that were similar to those seen in HRP-injected motoneurons, but their dendrites did not reach as far as the dendrites of the HRP-stained cells (1 mm for the radioactive glycin, and 300–800 mm for the Procion Yellow).

Thus, quantitative data on the distributions of dendrites and axons obtained with Goldi-impregnated material should be considered to underestimate the true figures. This book attempts, as far as possible, to use HRP-injected material for illustrating cortical cells.

9. Histochemical staining. Any specific enzymatic or chemical property that characterizes given neurons may be used to develop staining techniques that are specific to those neurons. For instance, it was recently discovered that if one compares the response properties of neurons situated in regions of the visual cortex that are rich in the enzyme cytochrome oxidase C with the properties of neurons situated in regions with little of this enzyme, one can discern clear differences. The former have round receptive fields and are sensitive to colors, whereas the latter have elongated receptive fields and are insensitive to colors [Livingstone and Hubel, 1984].

In recent years, because of great progress in molecular biology and immunological techniques, one can obtain antibodies for almost any macromolecule, and those antibodies can then be used to mark the cells containing a specific macromolecule. For instance, Schmechel et al. [1984] identified smooth stellate cells that contained glutamic acid decarboxylase (and presumably are using GABA as their inhibitory transmitter) and used immunocytochemistry to divide them into two subtypes according to the presence or absence of somatostatin.

10. Metabolic staining. Cells that are active metabolically may be marked with metabolic analogues. Sokoloff et al. [1977] introduced deoxyglucose (DOG) as such an analogue. This material is very similar to glucose, which is virtually the sole source of energy-providing metabolism of neurons. The DOG is absorbed and follows the initial metabolic processes as if it were glucose. However, once it becomes phosphorylated, it cannot be further metabolized, nor can it leave the cell. Thus, DOG-6-phosphate is trapped in the cells. Its rate of accumulation is proportional to the rate of glucose utilization by the cell. Presumably, cells that fire many spikes require more energy and accumulate the DOG-6-phosphate faster.

1.3.2 The organization of the cerebral cortex

Two types of cerebral cortex are commonly recognized, according to the relative placement of the white and gray matter. In the allocortex, the gray matter (containing most of the neurons and dendrites) is on the inner side, and the white matter (containing most of the myelinated axons) is on the outer side. In the isocortex, the order is reversed: The gray matter is on the outer side, and the white matter is inside. This classification is approximately the same as the archipallium/neopallium classification. The latter names were preferred by researchers who wished to emphasize the phylogenetic aspects of cortical evolution, and the allocortex/isocortex nomenclature was used by researchers wishing to stress the ontogenetic aspects of cortical development [Kappers, Hubers, and Crosby, 1936]. The isocortex accounts for most of the cortex (over 90 percent). From here on, we shall deal solely with it. When we say "cortex," we refer to the isocortex.

Most investigators today divide the cortex into six layers. According to Lorente de No [1949], this convention is based on a historical mistake. In 1909, Brodman described the development of the cortex in the human embryo. According to his description, up to the age of six months the cortex of the human embryo appeared homogeneous. At six months the cortex changed its appearance, and six layers could be observed. They appeared to be continuous and of equal width throughout the extent of the isocortex. Only later did the different layers show differential development in the various cortical areas. That description entrenched the idea that six layers represented the basic organization throughout the cortex. Later, when embryo cortex was stained by the Golgi impregnation technique, it became clear that the various cell types in the cortex could be seen before six months and that they were organized according to the adult pattern of layers, which does not overlap with the six layers observed by Brodman. Despite that discovery, the convention of dividing the cortex into six layers continued.

Although most investigators today are in agreement regarding the six-layer convention, the criteria according to which the layer borders are demarcated vary from one cortical area to another and from one researcher to another. One must take great care when attempting to compare descriptions of a given layer in different cortical areas or descriptions by different authors. The description given here is based on that of Lorente de No. The external layer of the cortex (called layer I) contains very few neurons. It is composed mostly of the terminal tufts of apical

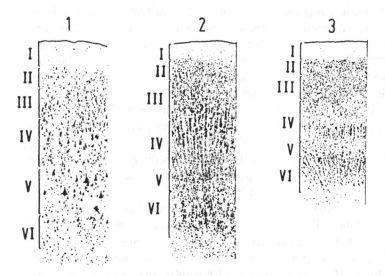

Figure 1.3.4. Nissl stains from three different cortical areas. [Based on A. W. Campbell, *Histological Studies on the Localization of Cerebral Function*, 1905, Cambridge University Press.]

dendrites of pyramidal cells and of a mesh of horizontal axons running in all directions. In the next layer (layer II), most of the cells are small pyramidal cells. In layer III, the pyramidal cells still prevail, but they are a little bit larger than those of layer II. Layer IV is relatively rich in stellate cells. In its upper part (called layer IVa) it contains a mixture of small and medium-size pyramidal cells and stellate cells, whereas in its deepest part (called layer IVb) it contains almost exclusively stellate cells. Layer V again contains a mixture of all types of cells, but it is differentiated from the other layers by the presence of very large pyramidal cells. Those are not present in layer VI. Many of the pyramidal cells of layer VI are elongated. Some elongated cell bodies have been called fusiform cells. The deepest part of layer VI is particularly rich with fusiform cells.

The class of fusiform cells was not mentioned in the description of cell types in Section 1.2 because it is not generally accepted today as a uniform class. It probably contains both deformed pyramidal cells and smooth stellate cells and perhaps some Martinotti cells as well. However, the fusiform shape is quite apparent in Nissl stains and can be used to characterize layer VI.

In Figure 1.3.4 we see three examples of cortical slices, taken from different cortical regions, stained in Nissl stain (which shows only the cell bodies). In example 2, the division into six layers is much clearer than in examples 1 and 3. It is customary to classify the cortex into "homotypical isocortex," in which the layers are clearly visible, and "heterotypical isocortex," in which not all the layers can be seen clearly. Example 1 in Figure 1.3.4 is taken from the primary motor cortex. In this heterotypical isocortex the layers of small cells (II and IVb) are poorly developed. This type of cortex is thus called "agranular." In contrast, these layers are overdeveloped in primary sensory cortices (example 3 in Figure 1.3.4). In these areas it is difficult to recognize the layers that contain middle-size pyramidal cells (III and IVa). These areas are called "granular cortex," or koniocortex.

Figure 1.3.4 shows that different cortical areas vary in the details of their appearance. These architectonic differences are used to divide the cortex into many subareas. At the turn of the twentieth century, many morphologists attempted to parcel the cortex according to various criteria. Of those, the parcelation most frequently used is the cytoarchitectonic parcelation of Brodman, who divided the human cortex into some fifty areas (Figure 1.3.5). The areas are numbered roughly in the order in which they are encountered as one proceeds from the top of the brain down. In many animals, homologous areas can be recognized, and they are designated by the same numbers. In several cases it was found in later years that what at first had been considered as one area should actually be divided into two subareas. That was done by adding a letter to the area's number (e.g., 3a and 3b in the somatosensory cortex).

Sometimes the cytoarchitectonic parcelation of von Bonin is used (Figure 1.3.6). This parcelation is less detailed than that of Brodman, and its nomenclature is derived from the place in which a region lies. For example, the frontal lobe is divided into FA, FB, FC, and so forth. Less popular are the cytoarchitectonic parcelations of von Economo (which is more detailed than the Brodman parcelation) and the myeloarchitectonic parcelation of Vogt (which yields results similar to those of Brodman).

The intensive concern with the parcelation of the cortex in the beginning of this century marked but another wave in the ongoing oscillation between holistic and compartmentalized views of cortical functions. These concepts can be traced back at least to the eighteenth century, when more and more scientists were accepting the idea that the brain is the organ of the mind. In the eighteenth century, the prominent scientist Albrecht von Haller (1708–77) argued that as the mind was one omnipo-

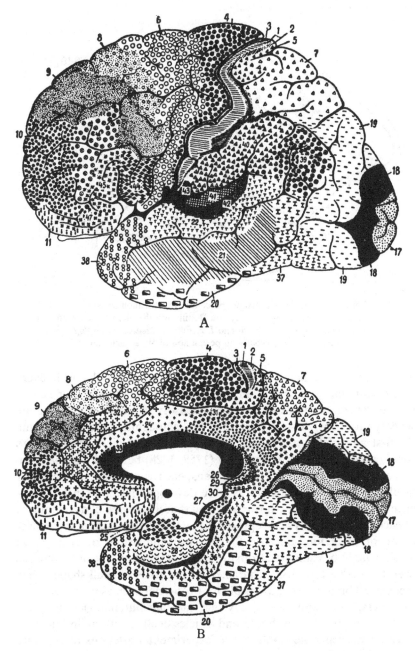

Figure 1.3.5. Cytoarchitectonic parcelation of Brodman. [From S. W. Ranson, *The Anatomy of the Nervous System from the Standpoint of Development and Function*, 1930, with permission of W. B. Saunders.]

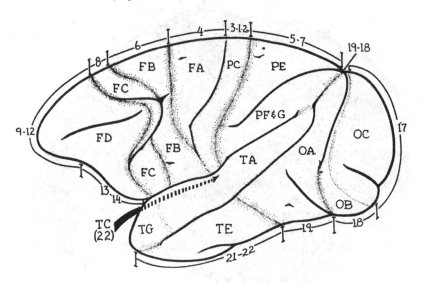

Figure 1.3.6. Cytoarchitectonic parcelation of the monkey brain. The letters indicate the names as given by von Bonin, and the digits were assigned by Brodman. [From T. C. Ruch and J. F. Fulton, *Medical Physiology and Biophysics*, 18th ed., 1960, with permission of W. B. Saunders.]

tent entity, so was the brain. Every piece of the brain took part in every type of mind process. It was only in the peripheral nerves that one could find specialization between motor and sensory functions and among the sensory functions. Toward the end of the eighteenth century and during the first half of the nineteenth century a different view became more widely accepted. Franz Joseph Gall (1758–1828) believed that the brain was actually a collection of many organs, each one representing another faculty of the mind. Those ideas were developed into a branch of "science" called phrenology. It was assumed that the degree of development of each mental faculty was represented by the physical size of the appropriate brain region. Because the physical dimensions of many brain regions might be estimated from the shape of the skull, the phrenologists tried to predict a person's character on the basis of skull shape. They recognized brain areas responsible for properties such as pride, bravery, smartness, love, and lust. Today one can find phrenological maps as decorations on covers of books and articles dealing with brain function.

Those concepts inevitably led to experimental attempts to generate mental deficits by destroying parts of the brain in experimental animals. However, lacking anesthesia and measures against infection, such tests were difficult to carry out. The first who succeeded to some extent was Marie Jean Pierre Flourens (1794–1867). With the crude surgical and observational methods available in his time he discovered that damage to the brain stem could result in specific sensory or motor deficits, whereas damage to higher brain structures resulted in generalized debilitating effects. Flourens seemed to prove that basic motor and sensory functions were seated in different regions of the brain stem, whereas more complex functions, such as sensory perception or the will to move, were distributed over all the higher brain regions. The concept of "higher brain functions" that is often used in modern neuroscience stems from the argumentation of that period. Flourens is said to have proved the omnipotence of the forebrain and the unity of the mind.

In the second half of the nineteenth century, with the improvement of experimental methodology, the thinking began to swing back toward the view that the cortical regions were specialized. In 1855, Panizza found that destruction of the occipital lobes caused blindness. In 1861, Broca found that damage to a small localized region in the brain of a patient resulted in disruption of the motor aspects of speech. In 1870, Hitzig showed that electrical stimulation of certain cortical regions could induce movements. Finally, Jean Bouillaud (1796–1881) produced maps showing the motor, somatosensory, visual, and auditory regions of the cortex. Bouillaud suggested that for every sensory (and motor) system there was a primary cortical area surrounded by a secondary sensory area and that in between them there were association areas. By the end of the nineteenth century and the beginning of the twentieth the concept of cortical specialization again was ascendant.

It is clear why parcelation of the cortex by architectonic principles seemed so relevant at that time. However, the intensive research into cortical architectonics gradually diminished when it was realized that mapping alone did not lead to an understanding of how the higher brain functions are carried out. Given the prevailing ideas on the parcelation of functions among cortical regions, it was natural that renewed efforts were made to identify, through the use of ablation methods, which region carried out which function. Probably the most prominent research of that type during the 1920s and 1930s was that of Karl Spencer Lashley (1890–1958), who attempted to discover where the learned trace was stored. That was done by training rats to discriminate between two shapes and then trying to "erase" that memory by controlled cortical ablations, or by cuts that separated cortical regions from each other. As long as the rats were not rendered blind by complete removal of the visual cortex, none of the lesions succeeded in erasing the memory. Following large lesions, the performance of the rats deteriorated, but the degree of deterioration was associated mainly with the amount of cortex ablated, rather than with which areas had been removed. These experiments were summarized by Lashley [1950] in his article "In Search of the Engram."

In another set of experiments, Lashley attempted to discover which brain areas were required for space orientation. Lashley [1929] trained rats to run mazes of varying complexity, and then he systematically removed brain regions in an effort to find where the engram of the maze was stored. The amount of behavioral deficit produced by the ablation was related to the amount of cortex removed, not to which part of the cortex was removed. Of course, particular deficits were found to be related to the removal of particular cortical areas: Rats from which the visual cortex was removed found their way in the maze by palpation, whereas those from which the motor cortex was removed had great difficulty in walking straight, and although they would crawl and roll over, they would take the correct turns and reach the goal. By comparing the memory and learning deficits produced by various cortical lesions, Lashley concluded that the deficit increased with the amount of cortex removed, and that more complex paradigms were more sensitive to damage.

The boundary between the specific and the omnipotent brain regions, which Haller had placed at the border between the peripheral and the central nervous systems, and Flourens had located between the brain stem and the forebrain, was said by Lashley to be situated between the specific (motor and sensory) cortices and the association cortex. Yet it should be noted that Lashley asserted that the specific cortices played roles in higher brain function, in addition to their specific roles. The idea that functions can be distributed across wide cortical areas has prompted many studies using statistical tools to evaluate cortical function and connectivity. The reader can gain much insight into the current debates by studying the behavioral work of Lashley, the anatomical work of Sholl [1956], the electrophysiological work of Burns [1958, 1968], and the multidisiplinary work of John [1967].

The notion that cortical functions were carried out through interactions among widespread areas was not accepted universally. Some of the strongest evidence against that view was provided by the brain-stimulation experiments of Penfield. Penfield treated cases of intractable epilepsy by ablating the cortical region in which the epileptic focus seemed to be located. In order to localize the epileptic center, Penfield operated on his patients under local anesthesia: A piece of skull was removed, the dura was reflected, and the exposed cortex was gently stimulated. The fully awake patient was asked to report the feeling aroused by the stimulus. Whereas stimulating the primary sensory areas produced only simple sensations of the appropriate modality, stimulating association areas could interfere with mental processes or even produce specific memories. By pooling his results from several patients in whom different brain areas had been exposed, Penfield produced a map of the functional roles of many areas of the cortex [Penfield and Rasmussen, 1950]. The map contains not only primary sensory and motor functions but also many higher functions, such as memory and speech.

In recent decades, the properties of single neurons have been measured in many different cortical areas by recording their spike activities. It has often been found that properties of neurons differ in different cytoarchitectonic areas. Improvements in tracing techniques have shown that different cytoarchitectonic areas also tend to differ in regard to the sets of structures from which they receive, and to which they send, connections. Refined behavioral testing techniques in many situations have succeeded in demonstrating specific aspects of deficits that are linked to ablations of certain cytoarchitectonic areas. It appears that the scientific pendulum is swinging once more toward the extreme position of dealing exlusively with the specific specializations of the cortical areas – so much so that in his introduction to a recent volume on the organization of the cerebral cortex, a prominent neuroscientist did not hesitate to refer to the work of Lashley as "behavioral experiments of rather dubious significance" and to the work of Sholl as "the sterile, statistical approach to cortical connectivity" [Cowan, 1981].

Despite the many detailed properties that can be used to differentiate among the various cortical areas, the common properties of all the cortical areas are overwhelming. The same cell types, the same types of connections, and the same distributions of cells and connections across the cortical depth are found in all the parts of the isocortex. These properties of the cortex are markedly different from those found in the other parts of the brain. It is reasonable to assume that the cortex can perform certain types of operations that other brain regions cannot. Each cortical region applies these operations to different types of data and produces different types of results according to the afferent and efferent connections of the region [Braitenberg, 1977]. Classification of cortical areas according to structural and functional differences cannot in itself bring about an understanding of the common operations conducted by the cortex and the manner in which they are carried out. Corticonics is the study of these common processes and their underlying neural mechanisms.

1.3.3 Connections with other brain regions

Every cortical region sends and receives connections to an from other areas through several routes: through the dense mesh of axons and dendrites in the border between it and its neighbors, through axons that run parallel to the cortical surface in layer I, and through the white matter below it.

When the cortical tissue is stained by the Nissl, Weigert, or Golgi method, its texture appears to be continuous, although in some cases with appropriate staining techniques one can observe discontinuities [LeVay, Hubel, and Wiesel, 1975]. Even the boundary between two cytoarchitectonic areas consists of a transition zone in which one can observe a gradual change from one texture to the other. Cortical neurons have dendritic trees that spread a fraction of a millimeter horizontally, as do the local axonal branches of the neighboring cells. It might seem that through these local branches and their connections information could spread tangentially across large distances. However, it has been shown that these connections are not involved in conducting information over long distances. Lashley [1929] found that several cuts across the cortex did not impair the performance or the learning ability of rats. Sperry [1947] made cuts across the somatosensory cortex in monkeys on a 3- by 3-mm grid and found no deficit as long as the white matter was not involved. The physiological evidence that properties of neurons tend to be organized in radial columns also contributed to the concept that the intracortical mesh of axonal and dendritic branches is used only for local processing of information, not for transmission across the cortex.

The axons in layer I spread up to distances of several millimeters. These axons are derived mostly from axons of Martinotti cells, from collaterals of axons that reach the cortex through the white matter, and to a lesser extent from the few neurons of layer I. An important role was attributed to these axons by Eccles [1981] and Braitenberg [1978]. However, there is no experimental evidence that can shed light on their role in information processing by the cortex. The experimental results of Sperry [1947] indicate that they are not involved in transmitting information over large distances. There is no doubt that most of the extrinsic cortical connections are carried through the white matter. These connections are of four types:

1. Connections between different cortical areas in one hemisphere
2. Connections between the two hemispheres
3. Connections between specific deep nuclei (mostly in the thalamus) and specific cortical areas
4. Diffuse connections between different areas of the brain stem and widespread areas of the cortex

All these connections are fundamentally mutual. That is, the cortex receives afferents from these regions and sends to them efferents.

Although the specific connections from thalamus to cortex, and back, have been studied most extensively, most of the connections in the white matter link the various cortical regions of the same hemisphere. All the pyramidal cells and some of the spiny stellate cells of the cortex send axons into the white matter. This means that about 80 percent of the cortical cells send axons into the white matter. In the human, there will be about 10^{10} axons in the white matter originating from the cortex. The corpus callosum (through which almost all the connections between the two hemispheres pass) contains "only" about 100 million axons. The cortex receives only 10–20 million axons from specific deep nuclei and sends back about the same number of axons. The number of unspecific axons is of a lower order of magnitude. Thus, about 98.6 percent of the axons that run in the white matter lead from one cortical region to another within the same hemisphere.

The axons that enter and leave the cortex tend to run in radial bundles. Between the bundles, the cell bodies of the neurons tend to be organized in radial chains. Close examination of cortical slices 2 and 3 in Figure 1.3.4 reveals these radial chains, which are prominent in layers VI, V, and IV. Within the cortex the afferent axons branch out and form many synaptic contacts; however, the density of branches and contacts is not homogeneous throughout the cortical depth. The corticocortical afferents tend to form connections in all the cortical layers, but those connections are more dense in the superficial layers. The specific afferents concentrate their terminals in the middle cortical layers, mostly in layer IV of the sensory cortices and in layer III of the association cortex [Jones, 1981]. The nonspecific axons innervate all the cortical layers, but at higher densities in layers VI and I. This pattern is illustrated in Figure 1.3.7.

In different cortical areas and in different animals there are numerous minor variations on this general scheme. For instance, in the primary visual cortex (area 17 of Brodman) of the cat, the specific thalamic

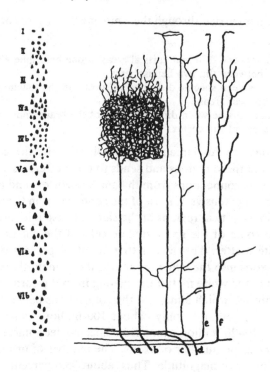

Figure 1.3.7. Afferent connections to the cortex: specific thalamic afferents (a, b), nonspecific afferents (c, d), corticocortical afferents (e, f). [Based on R. Lorente de No, "Cerebral Cortex: Architecture, Intracortical Connections, Motor Projections," in J. F. Fulton (ed.), *Physiology of the Nervous System*, 1949, with permission of Oxford University Press.]

afferents terminate not only in layer IV but also in the upper half of layer VI and in layer I [LeVay and Gilbert, 1976].

The sources of cortical efferents are not distributed homogeneously throughout the cortical depth either. Most of the corticocortical efferents originate in the pyramidal cells of layers II and III. The efferents that project back into the specific thalamic nuclei originate mostly in pyramidal cells in layer VI, and the large pyramidal cells of layer V specialize in sending their axons to the brain stem or to the spinal cord. It has been suggested that the main reason for the stratified appearance of the cortex is the segregation of afferent terminals and efferent cells in different depth profiles [Jones, 1981].

In the past it was thought that the afferents arriving at the cortex branched off to generate a continuous zone of termination across the cortical breadth. Hubel and Wiesel were the first to show conclusively

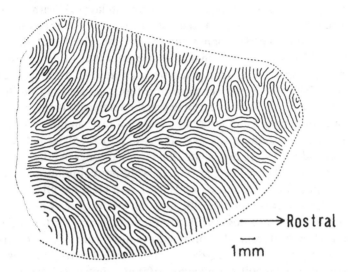

Figure 1.3.8. Parcelation of layer IV into zones that are affected only by one eye. The figure shows the full extent of area 17, which is located on the lateral surface of the left hemisphere. The black lines represent the lines that appeared pale in the silver staining. These lines mark the borders between areas that are affected by the left eye and areas that are affected by the right eye. [From S. LeVay, D. H. Hubel, and T. N. Wiesel, "The Pattern of Ocular Dominance Columns in Monkey Visual Cortex Revealed by Reduced Silver Stain," *Journal of Comparative Neurology,* 159:559–76, 1975, with permission of Alan R. Liss, Inc.]

that that was not the case. In their investigation of the response properties of neurons in the visual cortex they discovered that some neurons in layer IV would respond only to light patterns shone to the right eye, whereas others would respond only to patterns shone to the left eye, but they could find no neurons in layer IV that would respond to both. The neurons that responded to patterns shone to one eye (or the other) were not scattered randomly but were concentrated in alternating zones. These were called ocular dominance zones. Hubel, Wiesel, and their collaborators conducted numerous anatomical studies in which they showed that thalamic afferents (which bring information from the eyes) distributed their terminals in stripes that alternated between afferents bringing information from the left eye and those bringing information from the right eye. Finally, they discovered that in silver-stained cortical slices that were cut tangentially through layer IV, they could observe pale demarcation lines about 50 μm wide that separated the entire layer IV into two sets of convoluted stripes (Figure 1.3.8). By combining electrophysiological recordings from neurons in layer IV with histology, they succeeded in showing that the pale lines always marked the transition from one ocular dominance zone to another [LeVay et al., 1975).

These findings show that thalamic afferents that bring information from one eye tend to be segregated from afferents that bring information from the other eye. Each type of afferent gets a hold on its own territory in layer IV. The border between the territories is composed of a thin zone that contains no afferents (or only a few) and therefore looks pale in the silver stain. It is known that when the contacts between the thalamus and cortex are first made, the terminal zones overlap, and they become separated only later, as if there were mutual repulsion between the two types of terminals [LeVay, Wiesel, and Hubel, 1981]

A similar segregation of specific thalamic afferents was described by Woolsey and Van der Loos [1970] in the somatosensory cortex of the rat. There, thalamic afferents that bring information from different whiskers of the rat's snout end in segregated zones called barrels. It has been suggested that the picture of thalamic afferents shown by Lorente de No (Figure 1.3.7) is of such a barrel.

Similar nonrandom terminations of corticocortical afferents have been described by several authors. One of the earliest descriptions was that of Goldman and Nauta [1977], who applied radioactive amino acid to one cortical region and then followed the labeled axons to their termination zones. Figure 1.3.9 illustrates the termination zone around the principal sulcus (in the prefrontal cortex of the monkey) on one side, following application of the amino acid to the homologous zone of the other side. The black horizontal strip marks the arriving axons as they pass through the white matter. Some axons turn from the white matter to innervate radial patches of the cortex (columns).

The inverse relation can also be found. If one applies HRP to one small cortical region, one can expect to find the cells that become marked by the retrograde transport of the HRP segregated in small patches [Leichnetz, 1980]. Thus, cells from one cortical patch send their axons to several cortical patches, and each cortical patch receives cortical afferents from several cortical patches.

There is a marked correspondence between the division of the cortex into cytoarchitectonic areas and the division of the thalamus into nuclei. To a first approximation one can say that for every cortical area there exists a unique thalamic nucleus (or subnucleus) that sends to it afferents and receives from it efferents. In areas where the cortical surface can be mapped in greater detail (e.g., in sensory cortices), it has been shown

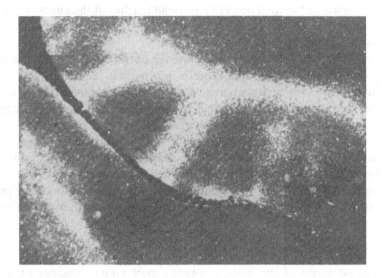

Figure 1.3.9. Distribution of corticocortical terminals. [From P. S. Goldman and W. J. H. Nauta, "Columnar Distribution of Cortico-cortical Fibers in the Frontal, Association, Limbic and Motor Cortex of the Developing Rhesus Monkey," *Brain Research,* 122:393–413, 1977, with permission of Elsevier Science Publishers.]

that the thalamocortical and the corticothalamic projections correspond. For example, the subregion in the thalamic nucleus that sends efferents to the cortex conveying information about touch sensation in the skin of the hand receives back projections from the hand area in the somatosensory cortex. Creutzfeldt [1978] suggested that the cytoarchitectonic differences are secondary expressions of the thalamocortical connectivity.

One must remember that most of the rules concerning the segregation of connections in the cortex do not represent absolute sharp boundaries, but are true only statistically. This point may be illustrated by examining in some detail the connections between the auditory cortex of the cat and the appropriate thalamic nucleus. The cortical auditory area in the cat may be divided into five different cytoarchitectonic regions. Along with this division it is possible to show, under the appropriate conditions, the existence of five maps of sensitivity to different tones. The thalamic inputs to these areas come from the medial geniculate nucleus of the thalamus, which also is divided into several subnuclei. It is normally stated that each cytoarchitectonic auditory area derives its input from a separate thalamic subnucleus. However, a close look at the results obtained by applying HRP to small patches of the auditory cortices (Figure 1.3.10) shows that to be true only if one considers the major source of input to each cortical area.

It is evident that complete knowledge of the ways in which the brain processes information will require information regarding the interactions between the cortical areas and the corresponding thalamic nuclei, the interactions among different cortical areas, and the local processes that take place inside every little patch of cortex. The objective of this book is to describe quantitatively the ways in which the neurons within the cortex interact locally. Before turning to that objective, in the next two sections we briefly consider the development of the cortex (which sheds light on many of the morphological features of the cortex previously discussed) and the quantitative aspects of cortical organization.

1.4 Embryogenesis of the cerebral cortex

The structure of the mature cortex shows a great number of details whose relevance to cortical function is unknown. It is often argued that if nature took the trouble to create the cortex in a certain way, it must have a value for the survival of the species. However, "survival value" does not necessarily imply better functioning in the adult years. Instead, it may mean a better chance for environmental adaptation at an early age, or a more efficient use of the genetic code that is required for the growth of the cortex in the embryo. Knowledge of the processes that lead to the formation of the cortex can help us assess the significance of the interal details of its organization. Therefore, this section briefly describes the development of the cortex in the mammalian embryo (and during the first postnatal days for some species).

The nervous system begins to develop in the human embryo at the age of three weeks, with the formation of the "neural plate." This is merely a thickened strip of the outer layer (ectoderm) of the embryo. As cells in the neural plate continue to grow and multiply at a high rate, the plate folds to form a groove, called the "neural groove." Finally, the neural groove closes off, forming the "neural tube," which becomes buried under the ectoderm. The entire central nervous system develops from the neural tube. The cells in the neural tube proliferate and grow unevenly, causing its anterior part to swell and bend so as to form three distinct parts: the hindbrain, the midbrain, and the forebrain. From the front end of the forebrain, two vesicles begin to develop, one on each side. These will form the neural part of the eyes. Slightly above the stalk

A

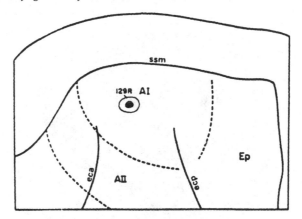

B

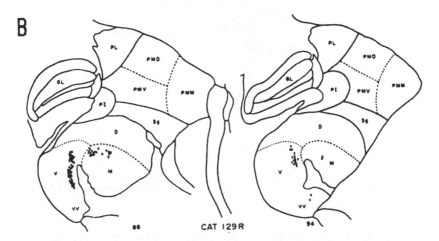

Figure 1.3.10. Retrograde labeling of medial geniculate body (MGB). A: Lateral view of the auditory cortex of the cat and its division into auditory fields. The dark spot marks the HRP injection site. The boundary of maximal local spread is marked around it. B: Two frontal sections through the thalamus. Places where labeled neurons were found are marked by ×. Although only a small spot in the center of AI was injected, the retrogradely labeled neurons were found in three subnuclei of the MGB. [Based on K. Niimi and H. Matsuoka, "Thalamocortical Organization of the Auditory System in the Cat Studied by Retrograde Axonal Transport of HRP," *Advances in Anatomy, Embryology and Cell Biology*, 57:1–56, 1979, with permission of Springer-Verlag.]

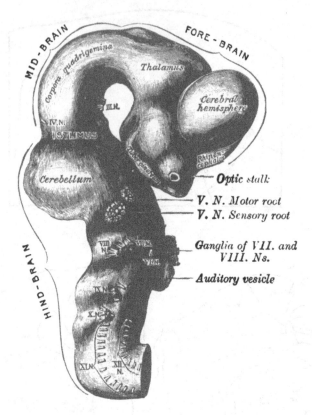

Figure 1.4.1. The brain of a 4.5-week-old human embryo. [From H. Gray, *Anatomy of the Human Body*, 26th ed., 1954, with permission of Lea & Febiger.]

of each eye vesicle, another vesicle starts to grow: the "telencephalic vesicle." The brain at this stage of development is shown in Figure 1.4.1. The lumen of the telencephalic vesicles will become the lateral ventricles, and the cerebral hemispheres will develop from the walls of these vesicles.

A common pattern of cell formation can be seen throughout the developing neural tube. The cells that multiply come from the innermost layers that line the lumen of the neural tube. Some of the newly formed cells remain near the lining layer and serve to enlarge the tube longitudinally and sideways, but the majority are pushed out into the walls of the tube. There they gradually become differentiated into two lines of cells: those that eventually become neurons (called neuroblasts) and those that develop into neuroglia. The cells of each brain region are generated in the same order. First the large neurons, then the small neurons, and then the neuroglia are generated [Jacobson, 1978]. However, the different types of cells are not generated sequentially, with an abrupt transition from one type to the other; rather, the cells' generation over time assumes the shape of a wave. The generating waves start successively overlapping each other to a large degree.

As this process continues, the thickness of the wall of the neural tube vastly increases. The newly formed cells near the tube's lumen, which are destined to be neurons, must travel long distances through the thickness of the tube wall before they reach their destination in the brain. There each neuroblast settles, grows an axon and dendrites, and becomes a neuron. The migration pattern of the neuroblasts determines much of the internal organization of many brain regions, including the cerebral cortex.

The process of thickening and differentiation starts earlier in the caudal parts of the neural tube. By the time the telencephalic vesicles begin to develop, the spinal cord, the hindbrain, and the midbrain contain many differentiated neurons. Even the deeper structures of the forebrain (such as the thalamus) have already reached an advanced stage of development. Many neurons in the lower brain levels send their axons into the outer walls of the growing telencephalic vesicles to form a thick felt of axons running tangentially in all directions.

The early afferent axons that arrive at the telencephalic vesicles seem to be monoaminergic [Molliver, 1982]. Their arrival marks the beginning of the growth of the telencephalic vesicles' wall into the cerebral cortex. According to Marin-Padilla [1984], the arrival of these afferent fibers triggers the differentiation of neuroblasts within the telencaphalic vesicles' wall into neurons. These differentiate into Cajal-Retzius cells, whose dendrites are the main targets of the early afferent axons and whose axons grow horizontally in all directions to form a second mesh of horizontal fibers. The differentiation of the Cajal-Retzius cells and the growth of their horizontal dendrites and axons seem to divide the fibrous walls of the telencephalic vesicles into two layers. The inner layer becomes the white matter, and the outer layer, including the Cajal-Retzius cells themselves, becomes layer I of the cerebral cortex.

When the telencephalic vesicles are formed, specialized glial cells bind the outer surface of the wall to its inner surface. As the wall's thickness is increased, both by the incoming fibers and by the local fibers, these glial cells remain anchored to the outer and inner surfaces. They become elongated and thin and form the "radial glial fibers." They play an important role in guiding the cortical neurons to their target positions, as described later.

According to Marin-Padilla [1984], the appearance of the Cajal-Retzius cells triggers the next stage of cortical differentiation. At this stage the cells that line the lumen of the telencephalic vesicles divide rapidly, generating large numbers of neuroblasts. The neuroblasts seem to be attracted to the layer of Cajal-Retzius cells, for they migrate until they reach that layer. On arrival, they grow dendrites into the layer of axons of the Cajal-Retzius cells and become cortical pyramidal cells. The immense numbers of neuroblasts that arrive form a dense layer of cells that completely separates the outer layer of fibers (layer I) from the inner layer of fibers (the white matter). This dense plate of neuroblasts and differentiating neurons is called the "cortical plate."

The migration path of the neuroblasts from the inner surface of the telencephalic vesicle, where they are formed, toward layer I, on the outer aspect of the wall, is not random. The neuroblasts seem to migrate along the radial glial fibers that connect the inner surface of the wall to its outer surface. Thus, according to Rakic [1982], all the pyramidal cells found in a vertical chain in the adult cortex were formed at a single point on the inner surface of the telencephalic vesicle. Within each vertical chain, the position of a pyramidal cell signifies its time of generation, the deeper cells having been born first.

This inverted order in depth is due to the continuous attraction of the newly formed neuroblasts to the layer of Cajal-Retzius cells. Thus, the neuroblasts that arrive at the cortical plate continue to push their way amidst the differentiated neurons until they reach the outer layer. There they anchor themselves to the fibers of layer I by growing dendrites into the layer. Through this process the newly arrived neuroblasts push the older ones downward. However, the older ones do not lose their anchorage to layer I. Rather, they develop the apical dendrite through which the cell body remains attached to its dendritic tuft in layer I. The last pyramidal neurons that are formed in this way are the pyramidal cells that are found in the outer part of layer II. These cells do not have the apical dendrite. Rather, they exhibit few dendrites that grow from the top of the cell body into layer I.

This description suggests that every pyramidal neuron has an apical dendrite that reaches layer I. According to Braitenberg [1978] and Marin-Padilla [1984], that is indeed the case. However, many other investigators have described pyramidal cells in the deep layers whose apical dendrites do not reach beyond layer III or even beyond layer IV. It is not clear whether these apparently short apical dendrites resulted from incomplete Golgi impregnation or from degeneration of apical dendrites that once reached layer I, or perhaps these cells arrived at the cortical plate at a late stage and failed to push all the way up to layer I.

The pyramidal cell grows an axon at an early stage. The axon always seem to grow downward, leaving the cortical plate and extending into the white matter, where it changes its course and runs horizontally. The local collaterals of the axons and the basal dendrites develop much later.

The Martinotti cells also start arriving at the cortical plate at an early stage, but in contrast to the pyramidal cells, their dendrites are not attracted to layer I. Instead, the Martinotti cell grows an axon in an upward direction. The axon enters layer I and runs in a horizontal direction, establishing synapses with the apical dendrites of the developing pyramidal cells. The apical dendritic tufts of the pyramidal cells develop earlier than the other dendrites. They receive many synapses in layer I, most of which probably are excitatory. There is a physiological need for some inhibitory feedback in this region in order to suppress the tendency of the excitatory network to explode into an epileptic fit (Chapter 5 will deal with this requirement extensively). This inhibitory feedback probably is supplied at an early stage both by the Martinotti cells, which are probably inhibitory [Marin-Padilla and Marin-Padilla, 1982], and by the newly arriving horizontal cells, which are GABA-ergic [Wolff et al., 1984].

The pyramidal neurons generated during cortical formation fill up the entire cortical plate. The nonpyramidal neurons seems to be generated in two stages. Shortly after the appearance of the pyramidal cells in the cortical plate, GABA-absorbing neuroblasts appear both in layer I and below the cortical plate. Then, after some delay, GABA-absorbing neuroblasts appear in the middle layers. Apparently, those appearing in the second stage are not attracted to any particular structure, but distribute themselves throughout the cortical depth [Wolff et al., 1984].

While the cortical plate is being formed, the thalamus is already fairly mature and contains many differentiated neurons. These send their axons toward the developing cortical plate; however, the axons stop their growth in the white matter underneath their target cortical zone [Rakic, 1984]. They remain there until all the neural components of the newly formed cortex reach their target positions. Only then do the thalamic axons resume their growth into the cortical plate and then branch off to form a dense plexus of axons in the middle of the cortex.

As mentioned earlier, the stellate cells tend to deposit themselves diffusely throughout the cortical plate. They appear in large numbers in the late stages of cortical development, and their main stage of dendritic and axonal growth is delayed until after the thalamic axons have grown into the cortical plate [Parnavelas and Ulings, 1980]. Pinto-Lord and Caviness [1979] suggested that the cell's dendrites grow by "attraction" to adjacent axonal tracts. In this way, stellate cells in the middle cortical layers, which are rich in horizontally oriented axons, grow dendrites in all directions and become multipolar stellate cells. The dendrites of stellate cells in superficial (or deep) cortical layers are attracted to fiber tracts below (or above). These stellate cells tend to have bipolar or bitufted shapes. The stellate cells are the only sources of inhibition inside the cortical gray matter. They arrive and form their inhibitory connections just as the local excitatory connections between the pyramidal cell axons and their dendrites begin to develop.

Not all areas of the cortex are formed simultaneously. First, only the primary sensory (and motor) areas are generated. At that time, the generation of new radial glial fibers (which guide the neuroblasts into their destination in the cortical plate) stops. Only after the primary areas have reached a fairly advanced stage do large numbers of new radial glial fibers start to reappear, guiding new neuroblasts into the association areas that are formed around each primary area [Rakic, 1984].

While all these processes take place, the extent of the newly formed cortex increases tremendously, but the total numbers of early afferent axons and of Cajal-Retzius cells do not increase. Thus, whereas these elements are packed very densely in the early stages, they become diluted as the cortex expands. Around the time of birth, many of the Cajal-Retzius cells degenerate. In the adult, some association cortices may contain no Cajal-Retzius cells at all. However, the horizontal axons of Cajal-Retzius cells can reach large distances, so that even in the adult all cortical regions will contain their axons in layer I [Marin-Padilla, 1984]. In the mature brain, over 90 percent of the neurons found in layer I are GABA-ergic [Hendry et al., 1987; Meinecke and Peters, 1987].

The immense lateral expansion of the cortex and the formation of folds (the gyri and the sulci) result in uneven movements of the deep layers, as compared with the superficial layers. This strain particularly affects the oldest pyramidal neurons; the cell body of such a neuron is in the deep layer, but it is anchored through its apical dendrite to layer I.

This explains why many of the pyramidal cells in layer VI tend to be tilted sideways and to have deformed, elongated cell bodies.

The final stages of cortical differentiation include rearrangement and adjustment of the finer details of cortical organization. This process extends in most mammals into postnatal age. Although some glial cells are found in the cortical plate at early stages [Rickmann and Wolff, 1985], the wave of generation of glial cells peaks as the surge of generation of neuroblasts fades. Glial cells also are formed from germinating cells that line the ventricular surface. They migrate and infiltrate the entire neopallium [Fujita, 1980]. In the white matter they differentiate into oligodendrocytes, which form a myelin sheath around many of the axons that pass there. In the gray matter they differentiate into the various forms of astrocytes, as well as into oligodendrocytes. This late appearance of the oligodendrocytes explains the late myelination of cortical axons. Early investigators of cortical histology (e.g., Cajal and Lorente de No) preferred to work with newborn animals because their incomplete myelination allowed the Golgi material to impregnate the axonal tree more thoroughly.

The radial glial fibers degenerate and disappear. However, their footprints remain in the form of the arrangement of the somata of cortical neurons in radially oriented strings. The neurons that were formed in large numbers try to establish and receive as many contacts as possible. Some do not quite make it and degenerate. Thus, at an early age there is a rapid decline in the number of neurons in the cortex.

The neurons that survive continue to elaborate their dendritic and axonal trees and to establish many more synaptic contacts. The dendritic spines are formed in large numbers at this late stage of maturation and seem to develop in excessive numbers before thinning out. The rapid growth in brain volume at an early age is not due to the formation of new neurons but to the growth and thickening of the processes of the existing neurons.

One of the best-known examples of the fine reorganization of connections that occurs in the course of cortical maturation is the rearrangement of the specific thalamic input to the primary visual cortex (area 17 of Brodman). As previously described, the axons from the thalamus approach the developing visual cortex and then remain in a waiting compartment until all the neuroblasts arrive at the cortical plate. They then resume their growth into the cortex and form a dense plexus of fibers in the middle cortical layers. At this stage the afferents that bring information from the two eyes into the cortex completely overlap with each other. Only later do the terminal branches become segregated into alternating stripes for the left eye and right eye, as described in Section 1.3.3 (Figure 1.3.8).

This process of forming stripes of connections is not limited to the specific thalamic afferents. When afferents from two or more cortical regions compete for the same cortical target area, the corticocortical terminals tend to be segregated in a similar process. Recent experiments by Constantine-Paton [1981], who implanted a third eye into an embryo of a frog, have suggested that this rearrangement of terminals from two similar sources is a general outcome of neural cooperation and competition.

The major destination of the optic nerve axons in the frog is a brain stem structure called the "optic tectum," where the axons terminate in an orderly retinotopical fashion. The normal frog has two tecta, one on each side. The axons from the two eyes cross over completely, so that the left eye innervates the right tectum, and the right eye innervates the left tectum. In the "three-eyed" frog, two of the eyes formed connections with one optic tectum. The retinotopic map was preserved, but the terminal branches of the axons from the two eyes were segregated into alternating stripes approximately 0.5 mm in width, much like the ocular dominance zones of the normal visual cortex. See von der Malsburg and Willshaw [1981] for further analysis of processes leading to the formation of cortical maps.

This is not the only case in which similar growth principles appear to hold for diverse brain structures. The development of the cerebellar cortex, for instance, shares many features with the cortical development. We can learn much about the rules that govern the growth and development of the brain by studying the cortical disorganizations in various mutants and comparing the effects that these mutations have on the organizations of other brain structures; see Caviness and Pearlman [1984] for such an analysis. Because several mutants exhibiting severe cortical malformations appear to possess normal cortical functions, one may assume that the features that are absent from these mutants are not essential for cortical function.

In the final stages of development, the connections may be modified according to the experience of the young animal. Here again the visual cortex supplies some of the best-known examples. LeVay et al. [1981] studied the development of the ocular dominance zones in the visual cortex of the monkey. They found that in the normal newborn monkey, the segregation between left-eye and right-eye dominance zones was very poor. These zones develop during the first few weeks of life. They found that if during that period the lids of one eye were sutured (preventing light patterns from reaching that retina), the ocular dominance zones of the other eye expanded at the expense of the zones of the sutured eye. But if the eyelids were sutured at the age of six weeks or later, the normal segregation of the terminals from the two eyes was not affected.

This chapter has shown that many features of the cortex are results of the processes that lead to its formation. Yet it is impossible to discern which features are merely by-products of these rules, and for which features nature "went out of its way" to assure their generation. Is the radial arrangement of neurons in the mature cortex merely a by-product of their guidance into the cortical plate by the radial glial fibers? Perhaps neurons that stem from a common germinating ancestor tend to form stronger synaptic contacts among themselves, and therefore the form of cell migration to the cortex dictates stronger interactions in the vertical direction. Or perhaps the mechanism of migration along radial glial fibers has developed because it is advantageous to have the cells of the mature cortex organized in vertical strings. There are no clear answers in these matters, but we should remain open to all the possibilities.

1.5 Quantitative considerations

This section deals with the cortical organization quantitatively. This aspect of cortical organization is neglected in many textbooks, but it is essential for any attempt to examine the plausibility of candidate cortical circuits. Therefore, we are concerned not only with the numbers, as cited in the literature, but also with the methods used in obtaining them.

1.5.1 The density of neurons in the cortex

Data on the densities of neurons in different cortical regions of various animals have been available for many years. Such data usually are based on neurons counted in slices stained with Nissl methods (described in Section 1.3.1). Because these methods stain the cell bodies of all the cells, and because nerve cells are quite distinguishable from glial cells, it is fairly easy to count the nerve cells seen in stained cortical slices.

Table 1.5.1 shows data from various sources regarding neuronal densities for different animals, for different cortical regions in a given animal,

Table 1.5.1. *Densities of neurons in the cortex (thousands per cubic millimeter)*

A		B		C	
Animal	Density	Region	Density	Layer	Density
Mouse	142.5	Visual	106	I	20
Rat	105.0	Somatosensory	60	II	82
Guinea pig	52.5	Auditory	43	III	62
Rabbit	43.8	Motor	30	IV	67
Cat	30.8			V	61
Dog	24.5			VI	77
Monkey	21.5				
Human	10.5				
Elephant	6.9				
Whale	6.8				

Notes: A: Neuronal densities in the motor cortex in various animals, based on
Tower and Elliot [1952] and Tower [1954]. B: Neuronal densities in various
cortical regions in the human, based on table 225 of Blinkov and Glezer [1968].
C: Neuronal densities in the various cortical layers of the visual cortex in the cat
[Sholl, 1956].

and for different layers in a given cortical region in one animal. A wide
range of values is presented, and the data are not always consistent. In
part A, we find that the neuronal density in the human motor cortex is
10,500 neurons per cubic millimeter, whereas in part B the density is
$30,000/mm^3$. Such discrepancies are not uncommon when comparing
data from different sources and may be attributed to methodological
differences. Two common sources of differences are discussed next.

When a brain is processed for histology, it is subject to shrinkage.
While the brain tissue is hardened by a fixing solution, it may lose water
and shrink. Then, if the brain is dehydrated, it may shrink again. The
process of embedding the brain causes additional shrinking, particularly
if the embedding process calls for heating the tissue. Usually one as-
sesses the degree of shrinkage by putting two marks on the fresh tissue,
measuring the distance between them, and then measuring the distance
again after completing the processing of the tissue. The linear shrinkage
factor for cortical tissue embedded in paraffin and stained with the Nissl
method was found to be 0.67–0.81 by Schuz and Palm [1989]. It is not
known if shrinking is homogeneous in all directions, but if it is, 1 mm^3 of
paraffin-embedded tissue corresponds to 2.9 ($1/0.7^3$) cubic millimeters
of fresh tissue.

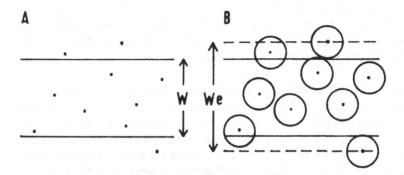

Figure 1.5.1. Correcting the slide thickness according to the size of the measured particles. A: For punctate objects, the thickness of the slide gives the true volume in which the particles are contained. B: For objects of finite size, the effective thickness is larger than the thickness of the slide.

Another source of error concerns the counting of neurons. Some neurons are cut on the plane of tissue sectioning. Parts of their cell bodies remain in the slide from which the cell count is taken, and the rest of their soma remain in the adjacent slide(s). Figure 1.5.1 depicts a block of tissue from which a slice of thickness W is cut. If the objects being counted were points (Figure 1.5.1A), then all the objects being counted would really be contained within this slice. However, if the objects are spheres of radius r (Figure 1.5.1B), then the centers of some spheres that can be seen in the slide are contained in a thicker slice. The effective thickness (W_e) can be computed [Abercrombie, 1946] as

$$W_e = W + 2r \qquad (1.5.1)$$

To apply this correction, we count only neurons whose nuclei can be seen in the slide. We also measure the average radius (r) of the neuronal nuclei. We then compute the effective thickness of the slide using equation (1.5.1) and divide the neuronal count by the effective volume of the slide.

The Abercrombie correction overestimates the effective thickness, because when only a small piece is left in the slide, it can no longer be recognized. Thus, instead of the object radius (r), we should take a smaller correcting distance ($r - b$), where b is the smallest thickness of the recognizable object. The effective width is given by

$$W_e = W + 2(r - b) \qquad (1.5.2)$$

Exercise 1.5.1. Are equations (1.5.1) and (1.5.2) accurate when the cells do not have homogeneous radii? Assume that the radii are distributed along some probability density function $f(r)$. How can one then correct for the finite size of the cells?

Despite the various sources of error that may arise while measuring neuronal densities, the following general properties of Table 1.5.1 remain valid: Small animals tend to have higher cell densities. The cortical thickness in animals with high cell densities is smaller; it is about 0.8 mm in the mouse and about 4 mm in the elephant (see Figure 1.1.3). It has been claimed that if one considered cylinders that were one cell wide and extended from the pia to the white matter, then one would find the same numbers of neurons in all mammals [Rockel, Hiorns, and Powell, 1980].

The cell densities in the various cytoarchitectonic areas of a given animal are not uniform. The motor cortex has the sparsest population, and sensory cortices tend to be populated more densely than the average. The primary visual cortex is the most densely populated area. The discrepancy is particularly noticeable in primates, where the neuronal density in the primary visual area is about twice that for other sensory cortices.

Even within a given cytoarchitectonic region, the cortical thickness and the neuronal density vary considerably. In the depth of the sulci, the cortex becomes thinner, whereas at the summit of the gyri it becomes thicker. Even the relative thicknesses of the various cortical layers change from the sulci to the gyri.

Despite all these differences, we believe that all cortical regions in all animals process information locally according to the same principles. Thus, the exact figures are not crucial for understanding these principles. In the following discussion we make frequent use of "typical" figures, assuming the cortical thickness to be 2 mm and the neuronal density to be 40,000 per cubic millimeter.

1.5.2 Relative abundances of cell types

Measurements of the relative numbers of cell types usually are based on Golgi-impregnated material, which shows the dendritic arborization and the presence of dendritic spines. For that reason, data are available mostly for classifications that use dendritic arborization as their criterion (as opposed to axonal arborization).

Previously it had been accepted that approximately 70 percent of the

Table 1.5.2. *Distributions of neuron types in various animals and various regions*

Animal	Region	Pyramidal	Stellate	Fusiform
Cat[a]	Visual	62%	34%	4%
	Somatosensory	63%	35%	2%
	Motor	85%	10%	5%
Monkey[a]	Visual	52%	46%	2%
	Motor	74%	22%	4%
Human[a]	Prefrontal	72%	26%	2%

Animal	Region	Pyramidal	Smooth stellate	Others
Rabbit[b]	Auditory	86.7%	9.5%	3.8%
Rat[c]	Visual II + III	87%		
	IV	90%		
	V	89%		
	VIa	97%		

[a]Data from Mitra [1955]; stellate cells here include both smooth and spiny stellates. Fusiforms are found mostly in the deep parts of layer VI.
[b]Data from McMullen, Glaser, and Tagaments [1984].
[c]Data from Peters and Kara [1985b]; the deepest part of layer VI was not included. The Roman numerals represent the layers within the primary visual cortex.

cortical cells were pyramidal cells, but in recent years there has been a growing tendency among several investigators to include most of the spiny cells in one category with the pyramidals. If one wishes to make but a crude division between (presumably) excitatory and inhibitory neurons in the cortex, it seems that 85–90 percent are excitatory, and only 10–15 percent are inhibitory. Table 1.5.2 and Figure 1.5.2 show the data on which this division is based.

When cell counts are conducted on Golgi-impregnated material, one assumes that all cell types have equal chances of being impregnated. As discussed in Section 1.3.1, it is not clear that that is exactly the case for all varieties of Golgi methods. The following hypothetical exercises give some sense of how a bias in the selectiveness of the staining method may affect the results.

Exercise 1.5.2. In a certain cortical region it was found that 75 percent of the neurons were pyramidal cells, 15 percent were smooth

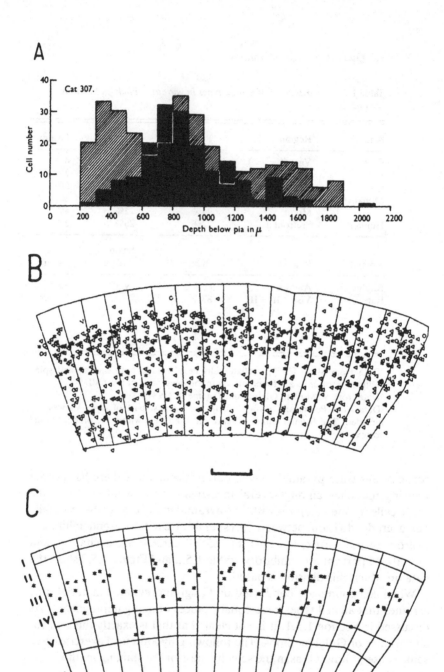

Figure 1.5.2. Distributions of neuronal types in the various cortical layers. A:
Hatched areas show numbers of pyramidal cells; black areas show numbers
of stellate cells (including spiny stellates). Note that despite the uneven

D

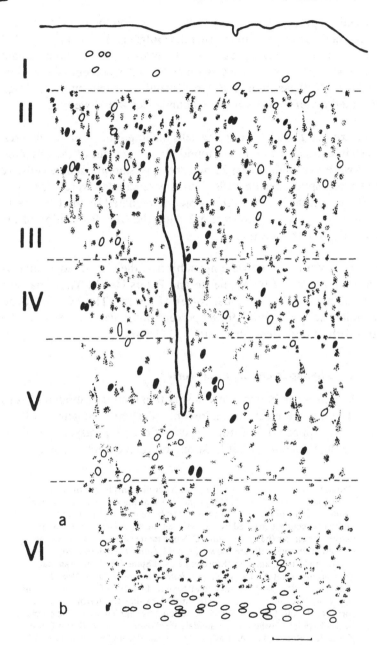

I

II

III

IV

V

a

VI

b

Caption to Figure 1.5.2 *(cont.)* distribution, there are no sharp demarcation lines between the cortical layers. [From N. L. Mitra, "Quantitative Analysis of Cell Types in Mammalian Neocortex," *Journal of Anatomy,* 89:467–83,

stellate cells, 8 percent were spiny stellate cells, and the remainder were Martinotti cells. It was also found that the Golgi precipitate impregnated only 2 percent of the pyramidal cells, 3 percent of the smooth stellate cells, 1.5 percent of the spiny stellate cells, and 1 percent of the Martinotti cells. If these four cell types could be seen only in Golgi-impregnated material, what relative abundances would we have?

Exercise 1.5.3. By counting cells in Golgi-impregnated material in a certain cortical region it was found that 60 percent of the observed neurons were pyramidal cells, 24 percent were smooth stellate cells, 12 percent were spiny stellate cells, and the remainder were Martinotti cells. The percentages of cells that were impregnated by the Golgi technique were as given in Exercise 1.5.2. Find the true distribution of the neuronal types in that cortex.

One of the few studies that were not subject to the bias of Golgi impregnation was carried out by Meinecke and Peters [1987]. They analyzed some 1,100 neurons in the visual cortex of the rat. The pyramidal and spiny stellate cells accounted for 85 percent of the neurons. Most of the neurons with smooth dendrites contained GABA.

1.5.3 Synaptic density in the cortex

Synapses can be seen only in very thin sections (about 0.05 μm) examined under the electron microscope. They are recognized by the characteristic apposition of the presynaptic and postsynaptic membranes, by the thickening of at least one of the membranes, and by the

Caption to Figure 1.5.2 *(cont.)* 1955, with permission.] B: All the neurons impregnated by the Golgi technique in a 0.3-mm-thick slide taken from the auditory cortex of the rabbit. Pyramidal cells are marked by triangles, smooth multipolar cells are marked by stars, and all the other nonpyramidals are marked by circles (bar = 0.5 mm). C: Same material as in B, but only the smooth multipolar cells are shown. [B and C from N. T. McMullen, E. M. Glaser, and M. Tagaments, "Morphometry of Spine-free Nonpyramidal Neurons in Rabbit Auditory Cortex," *Journal of Comparative Neurology*, 222:383–95, 1984, with permission of Alan R. Liss, Inc.] D: All the neurons whose nuclei could be seen in a 0.001-mm-thick slide taken from the visual cortex of the rat. Pyramidal cells are outlined by stippled areas, and bipolar cells by black spots; all other nonpyramidals are shown by outline (bar = 0.1 mm). [From A. Peters and D. A. Kara, "The Neuronal Composition of Area 17 of the Rat Visual Cortex. II. The Nonpyramidal Cells," *Journal of Comparative Neurology*, 234:242–63, 1985, with permission of Alan R. Liss, Inc.]

Table 1.5.3. *Synaptic densities in various animals*

Animal	Age	Region	Density ($\times 10^9$/mm^3)	Diameter (μm)
Human[a]	Adult		0.6	0.3
Human[b]	16–72 years	Frontal	1.1	0.24
Human[b]	74–90 years	Frontal	0.96	0.24
Cat[c]	5 weeks	Visual	0.79	0.26
Rabbit[d]	Adult	Visual	0.7	
		Motor	0.67	
Mouse[e]	4–6 months	Visual	0.75–0.81	0.27
		Premotor	0.67–0.74	0.30
		Frontal	0.64–0.71	0.31

Note: Synaptic diameter is the average length of the observed dense synaptic membranes.
[a]Data from Cragg [1975a], averaged from pieces of normal cortex that were excised during brain surgery in adult men.
[b]Data from Huttenlocher [1979], who used the Bloom and Aghajanian staining method.
[c]Data from Cragg [1975b].
[d]Data from Vrensen, De Groot, and Nuñez-Cardozo [1977].
[e]Data from Schuz and Palm [1989].

presence of synaptic vesicles. There are methods that seem to stain only the synaptic junctions [Bloom and Aghajanian, 1968]. Such methods could facilitate measurement of synaptic densities, but they have not been widely used for that purpose in the past.

Computing densities from counts of profiles in thin sections is subject to numerous systematic errors. In recent years, new methods for measuring such densities have been developed. These methods are collectively known as "stereology." A useful summary of such methods, with examples and mathematical justification, can be found in the work of Weibel [1979, 1980].

Problem 1.5.1. Corrections made for synaptic densities assume that the synapses are spheres. However, the counted structures are the dense synaptic membranes, which are better described as discs. How can one correct for counting disc-like objects? What is the error in an estimate caused by correcting for spheres? Consult the sources cited in Table 1.5.3, and evaluate the bias that their estimates may have had.

Table 1.5.3 shows some of the figures for synaptic density that have been published. The similarities between figures for different animals, different brain regions, and different investigators are amazing, particularly if compared with the large variations in neuronal densities (Table 1.5.1). This suggests that the limiting factor for local cortical connectivity is the packing density of the synapses. All cortices have synapses that are packed as densely as possible. Different degrees of local connectivities are achieved by changes in the number of dendrites that each neuron possesses.

If we assume a synaptic density of $8 \cdot 10^8$ per 1 mm^3 in all cortices, then in a mouse (with 100,000 neurons per 1 mm^3), each neuron receives 8,000 synapses. In a monkey (with 40,000 neurons per 1 mm^3), each neuron receives 20,000 synapses, and in a human (with 20,000 neurons per 1 mm^3) each neuron receives 40,000 synapses.

According to Braitenberg [1978], a synapse is formed, on the average, every 0.5 μm of dendrite. That means that in 1 mm^3 of cortex there are about 400 m of dendrites. Thus, in the mouse, each neuron has, on the average, 4 mm of dendrites, whereas in the human each neuron has about 20 mm of dendrites. Similarly, if one assumes that in the cortex an axon makes a synapse every 4 μm, then there are about 3,200 m of axons per 1 mm^3 of cortex. These estimates are very crude. Length densities can be measured quite accurately and easily by using the appropriate stereological methods [Weibel, 1979].

About 10 percent of synapses have symmetric profiles [Fisken, Garey, and Powell, 1975; Schuz and Palm, 1989]. Presumably these are the inhibitory synapses generated by local smooth stellate cells. Because the smooth stellate cells are concentrated in layer IV, it is not surprising that the symmetric synapses are also more concentrated in layer IV; according to White [1989], about 20 percent of the synaptic profiles in layer IV are symmetric.

Summed over the entire cortex, almost all synapses are generated by cortical neurons. Thus, each neuron in the mouse generates about 8,000 synapses, whereas in the human each cortical neuron generates about 40,000 synapses. However, one must remember that not all these synapses are generated locally. For most of the cortical neurons (the pyramidal cells), the neuron sends its main axon, or axonal collaterals, to other cortical areas as well. A rough estimate is that about half of the synapses are derived from axons that reach the cortex through the white matter, and the other half form local circuits. The remainder of this book is devoted to quantitative analysis of these local circuits.

Table 1.5.4 summarizes the quantitative data for a sort of "average" typical cortex. We use these figures as needed in the following chapters.

Table 1.5.4. *Typical compositions of cortical tissues*

Variable	Value
Neuronal density	40,000/mm^3
Neuronal composition:	
Pyramidal	75%
Smooth stellate	15%
Spiny stellate	10%
Synaptic density	$8 \cdot 10^8$/mm^3
Axonal length density	3,200 m/mm^3
Dendritic length density	400 m/mm^3
Synapses per neuron	20,000
Inhibitory synapses per neuron	2,000
Excitatory synapses from remote sources per neuron	9,000
Excitatory synapses from local sources per neuron	9,000
Dendritic length per neuron	10 mm

More extensive data of this type, for the cortex of the mouse, can be found in the work of Braitenberg [1986] and Schuz and Palm [1989].

1.6 References

Abercrombie M. (1946). Estimation of nuclear population from microtome sections. *Anat. Rec.* 94:239–47.

Blinkov S. M. and Glezer I. I. (1968). *Human Brain in Figures and Tables: A Quantitative Handbook* (translated by B. Haigh). Plenum Press, New York.

Bloom F. F. and Aghajanian G. K. (1968). Fine structural and cytochemical analysis of the staining of synaptic junctions with phosphotungstic acid. *J. Ultrastruct. Res.* 22:361–75.

Braitenberg V. (1977). *On the Texture of Brains.* Springer, Berlin.

(1978). Cortical architectonics: General and areal. In Brazier M. A. B. and Petche M. (eds.), *Architectonics of the Cerebral Cortex*, pp. 443–65. Raven Press, New York.

(1986). Two views of the cerebral cortex. In Palm G. and Aertsen A. (eds.); *Brain Theory*, pp. 81–96. Springer, Berlin.

Brown A. G. and Fyffe R. E. W. (1981). Direct observations on the contacts made between Ia afferent fibers and α-motoneurons in the cat's lumbosacral spinal cord. *J. Physiol. (Lond.)* 313:121–40.

Burns B. D. (1958). *The Mammalian Cerebral Cortex. Monographs of the Physiological Society, Vol. 8.* Edward Arnold, London.

(1968). *The Uncertain Nervous System.* Edward Arnold, London.

Campbell A. W. (1905). *Histological Studies on the Localization of Cerebral Function.* Cambridge University Press.

Caviness V. S. and Pearlman A. L. (1984). Mutation-induced disorders of mammalian forebrain development. In Sharma S. C. (ed.), *Organizing Principles of Neural Development,* pp. 277–305. Plenum Press, New York.

Colonnier M. (1981). The electron-microscopic analysis of the neuronal organization of the cerebral cortex. In Schmitt F. O., Worden F. G., Adelman G., and Dennis S. G. (eds.), *The Organization of the Cerebral Cortex,* pp. 125–152. MIT Press, Cambridge, Mass.

Constantine-Paton M. (1981). Induced ocular-dominance zones in tectal cortex. In Schmitt F. O., Worden F. G., Adelman G., and Dennis S. G. (eds.), *The Organization of the Cerebral Cortex,* pp. 47–67. MIT Press, Cambridge, Mass.

Cowan W. M. (1981). Keynote. In Schmitt F. O., Worden F. G., Adelman G., and Dennis S. G. (eds.), *The Organization of the Cerebral Cortex,* pp. XI–XXI. MIT Press, Cambridge, Mass.

Cragg B. G. (1975a). The density of synapses and neurons in normal, mentally defective and aging human brains. *Brain* 98:81–90.

(1975b). The development of synapses in kitten visual cortex during visual deprivation. *Exp. Neurol.* 46:445–51.

Creutzfeldt O. D. (1978). The neocortical link: Thoughts on the generality of structure and function of the neocortex. In Brazier M. A. B. and Petche M. (eds.), *Architectonics of the Cerebral Cortex,* pp. 357–83. Raven Press, New York.

Eccles J. C. (1981). The modular operation of the cerebral neocortex considered as a material basis of mental events. *Neuroscience* 6:1839–56.

Feldman M. L. (1984). Morphology of the neocortical pyramidal neuron. In Peters A. and Jones E. G. (eds.); *Cerebral Cortex, Vol. 1,* pp. 123–200. Plenum Press, New York.

Fisken R. A., Garey L. J., and Powell T. P. S. (1975). The intrinsic, association and commissural connections of area 17 of the visual cortex. *Philos. Trans. R. Soc. Lond. [Biol.]* 272:487–536.

Freund T. F., Martin K. A., Smith A. D., and Somogyi P. (1983). Glutamate decarboxylase immunoreactive terminals of Golgi impregnated axoaxonic and of presumably basket cells in synaptic contact with pyramidal neurons of the cat's visual cortex. *J. Comp. Neurol.* 221:263–78.

Friedlander M. J., Lin C. S., Stanford C. R., and Sherman S. M. (1981). Morphology of functionally identified neurons in lateral geniculate nucleus of the cat. *J. Neurophysiol.* 46:80–129.

Fujita S. (1980). Cytogenesis and pathology of neuroglia and microglia. *Path. Res. Pract.* 168:271–8.

Gerfen C. R. and Sawchenko P. E. (1984). An anterograde neuroanatomical tracing method that shows the detailed morphology of neurons, their axons and terminals: Immunohistochemical localization of an axonally transported plant lectin, *Phaseolus vulgaris* leucoagglutinin (PHA-L). *Brain Res.* 290:219–85.

Gilbert C. D. and Wiesel T. M. (1979). Morphology and intracortical projection of functionally characterized neurons in the cat visual cortex. *Nature (Lond.)* 280:120–5.

(1981). Laminar specialization and intracortical connections in cat primary visual cortex. In Schmitt F. O., Worden F. G., Adelman G., and Dennis S. G. (eds.), *The Organization of the Cerebral Cortex*, pp. 163–91. MIT Press, Cambridge, Mass.

Goldman P. S. and Nauta W. J. H. (1977). Columnar distribution of cortico-cortical fibers in the frontal, association, limbic and motor cortex of the developing rhesus monkey. *Brain Res.* 122:393–413.

Gray E. G. (1959). Axosomatic and axodendritic synapses in the cerebral cortex: An electron microscopic study. *J. Anat.* 93:420–33.

Gray H. (1954). *Anatomy of the Human Body*, 26th ed. Lea & Febiger, Philadelphia.

Hendry S. H. C., Schwark H. D., Jones E. G., and Yan J. (1987). Numbers and proportions of GABA-immunoreactive neurons in different areas of monkey cerebral cortex. *J. Neurosci.* 7:1503–19.

Homos J. E., Davis T. L., and Sterling P. (1983). Four types of neurons in layer IVab of cat cortical area 17 accumulate ^{3}H-GABA. *J. Comp. Neurol.* 217:449–57.

Huttenlocher P. R. (1979). Synaptic density in human frontal cortex – developmental changes and effects of aging. *Brain Res.* 163:185–205.

Hyden H. (1967). Dynamic aspects of the neuron–glia relationship – a study with microchemical methods. In Hyden H. (ed.), *The Neuron*, pp. 179–216. Elsevier, Amsterdam.

Jacobson M. (1978). *Developmental Neurobiology*, 2nd ed. Plenum Press, New York.

John E. R. (1967). *Mechanisms of Memory*. Academic Press, New York.

Jones E. G. (1981). Anatomy of the cerebral cortex: Columnar input–output organization. In Schmitt F. O., Worden F. G., Adelman G., and Dennis S. G. (eds.), *The Organization of the Cerebral Cortex*, pp. 199–235. MIT Press, Cambridge, Mass.

Kappers C. U. A., Huber G. C., and Crosby E. C. (1936). *The Comparative Anatomy of the Nervous System of Vertebrates, Including Man.* Macmillan, New York.

Kelly J. P. and Van Essen D. C. (1974). Cell structure and function in the visual cortex of the cat. *J. Physiol. (Lond.)* 238:515–47.

Kuypers H. G. J. M., Bentivoglio M., Castman-Berrevoets C. E. and Bharos A. T. (1980). Double retrograde neuronal labeling through divergent axon collaterals, using two fluorescent tracers with the same excitation wavelength which label different features of the cell. *Exp. Brain Res.* 40:383–92.

Lashley K. S. (1929). *Brain Mechanisms and Intelligence.* Chicago University Press.

(1950). In search of the engram. *Symp. Soc. Exp. Biol.* 4:478–80.

Leichnetz G. R. (1980). An intrahemispheric columnar projection between two cortical multisensory convergence areas (inferior parietal lobule and prefrontal cortex): An anterograde study in macaque using HRP gel. *Neurosci. Lett.* 18:119–24.

LeVay S. and Gilbert C. D. (1976). Laminar patterns of geniculocortical projection in the cat. *Brain Res.* 113:1–20.

LeVay S., Hubel D. H., and Wiesel T. N. (1975). The pattern of ocular dominance columns in monkey visual cortex revealed by reduced silver stain. *J. Comp. Neurol.* 159:559–76.

LeVay S., Wiesel T. N., and Hubel D. H. (1981). The postnatal development and plasticity of ocular-dominance columns in the monkey. In Schmitt F. O., Worden F. G., Adelman G., and Dennis S. G. (eds.), *The Organization of the Cerebral Cortex,* pp. 29–45. MIT Press, Cambridge, Mass.

Livingston M. S. and Hubel D. H. (1984). Anatomy and physiology of a color system in the primate visual cortex. *J. Neurosci.* 4:309–56.

Lorente de No R. (1933). Studies on the structure of the cerebral cortex. *J. Psychol. Neurol.* 45:382–438.

    (1949). Cerebral cortex: Architecture, intracortical connections, motor projections. In Fulton J. F. (ed.), *Physiology of the Nervous System,* pp. 268–312. Oxford University Press.

McMullen N. T., Glaser E. M., and Tagaments M. (1984). Morphometry of spine-free nonpyramidal neurons in rabbit auditory cortex. *J. Comp. Neurol.* 222:383–95.

Marin-Padilla M. (1984). Neurons of layer I: A developmental analysis. In Peters A. and Jones E. G. (eds.), *Cerebral Cortex, Vol. 1,* pp. 447–78. Plenum Press, New York.

Marin-Padilla M. and Marin-Padilla M. T. (1982). Origin, prenatal development and structural organization of layer I of the human cerebral (motor) cortex: A Golgi study. *Anat. Embriol.* 164:161–206.

Martin K. A. and Whitteridge D. (1984). Form, function and intracortical projection of spiny neurons in the striate visual cortex of the cat. *J. Physiol. (Lond.)* 353:463–504.

Meinecke D. L. and Peters A. (1987). GABA immunoreactive neurons in rat visual cortex. *J. Comp. Neurol.* 261:388–401.

Mesulam M. M. 1978). Tetramethyl benzidine for horseradish peroxidase neurohistochemistry: A noncarcinogenic blue reaction product with superior sensitivity for visualizing neural afferents and efferents. *J. Histochem. Cytochem.* 26:106–17.

Meyer G. (1983). Axonal patterns and topography of short-axon neurons in visual areas 17, 18 and 19 of the cat. *J. Comp. Neurol.* 220:405–38.

Mitra N. L. (1955). Quantitative analysis of cell types in mammalian neocortex. *J. Anat.* 89:467–83.

Molliver M. E. (1982). Monoamines in the development of the cortex. In Rakic P. and Goldman-Rakic P. S. (eds.), *Development and Modifiability of the Cerebral Cortex,* pp. 492–507. MIT Press, Cambridge, Mass.

Niimi K. and Matsuoka H. (1979). Thalamocortical organization of the auditory system in the cat studied by retrograde axonal transport of HRP. *Adv. Anat. Embryol. Cell Biol.* 57:1–56.

Parnavelas J. G. and Ulings H. B. M. (1980). The growth of non-pyramidal neurons in the visual cortex of the rat: A morphometric study. *Brain Res.* 193:373–82.

Pasternak J. F. and Woolsey T. A. (1975). On the selectivity of the Golgi Cox method. *J. Comp. Neurol.* 160:307–12.

Penfield W. and Rasmussen T. (1950). *The Cerebral Cortex of Man: A Clinical Study of Localization of Function.* Macmillan, New York.

Peters A. and Jones E. G. (eds.) (1984). *The Cerebral Cortrex, Vol. 1.* Plenum Press, New York.

Peters A. and Kaiserman-Abramof I. R. (1970). The small pyramidal neuron of the rat cerebral cortex. The perikaryon, dendrites and spines. *Am. J. Anat.* 127:321–56.

Peters A. and Kara D. A. (1985a). The neuronal composition of area 17 of the rat visual cortex. I. The pyramidal cells. *J. Comp. Neurol.* 234:218–41.

(1985b). The neuronal composition of area 17 of the rat visual cortex. II. The nonpyramidal cells. *J. Comp. Neurol.* 234:242–63.

Pinto-Lord M. C. and Caviness V. S. Jr. (1979). Determinants of cell shape and orientation: A comparative Golgi analysis of cell–axon interrelationships in the developing neocortex of normal and reeler mice. *J. Comp. Neurol.* 187:49–69.

Rakic P. (1982). Developmental events leading to laminar and areal organization of the neocortex. In Schmitt F. O., Worden F. G., Adelman G., and Dennis S. G. (eds.), *The Organization of the Cerebral Cortex,* pp. 6–28. MIT Press, Cambridge, Mass.

(1984). Organizing principles for development of primate cerebral cortex. In Sharma S. C. (ed.), *Organizing Principles of Neural Development,* pp. 21–48. Plenum Press, New York.

Ranson S. W. (1930). *The Anatomy of the Nervous System from the Standpoint of Development and Function.* Saunders, Philadelphia.

Rickmann M. and Wolff J. R. (1985). Prenatal gliogenesis in the neopallium of the rat. *Adv. Anat. Embryol. Cell Biol.* 93:1–104.

Rockel A. J., Hiorns R. W., and Powell T. P. S. (1980). The basic uniformity of the neocortex. *Brain* 103:221–44.

Ruch T. C. and Fulton J. F. (1960) *Medical Physiology and Biophysics,* 18th ed. Saunders, Philadelphia.

Schmechel D. E., Vickrey B. G., Fitzpatrick D., and Elde R. P. (1984). GABAergic neurons of mammalian cerebral cortex: Widespread subclass defined by somatostatin content. *Neurosci. Lett.* 47:227–32.

Schuz A. and Palm G. (1989). Density of neurons and synapses in the cerebral cortex of the mouse. *J. Comp. Neurol.* 286:442–55.

Sholl D. A. (1956). *The Organization of the Cerebral Cortex.* Methuen, London.

Sokoloff L., Reivich M., Kennedy C., Des Rosiers M. H., Pattak C. S., Pettigrew K. D., Sakurada O., and Shinohara M. (1977). The [C^{14}]deoxyglucose method for the measurement of local cerebral glucose utilization: Theory, procedure and normal values in the conscious and anesthetized albino rat. *J. Neurochem.* 28:897–916.

Somogyi P. (1977). A specific "axo-axonal" interneuron in the visual cortex of the rat. *Brain Res.* 136:346–50.

Somogyi P., Kisvarday Z. F., Martin K. A. C., and Whitteridge D. (1983). Synaptic connections of morphologically identified and physiologically

characterized large basket cells in the striate cortex of the cat.
Neuroscience 10:261–94.

Sperry R. W. (1947). Cerebral regulation of motor coordination in monkeys
following multiple transections of sensorimotor cortex. *J. Neurophysiol.*
10:275–94.

Tower D. B. (1954). Structural and functional organization of mammalian
cerebral cortex: The correlation of neuron density with brain size. *J.
Comp. Neurol.* 101:9–52.

Tower D. B. and Elliott K. A. C. (1952). Activity of the acetyl-choline
system in cerebral cortex of various unanesthetized animals. *Am. J.
Physiol.* 168:747–59.

Van der Loos H. (1967). The history of the neuron. In Hyden H. (ed.), *The
Neuron*, pp. 1–47. Elsevier, Amsterdam.

von der Malsburg C. and Willshaw D. (1981). Co-operativity and brain
organization. *TINS* 4:80–3.

Vrensen G., Cardozo N., Muller L., and Van der Want J. (1980). The
presynaptic grid: A new approach. *Brain Res.* 184:23–40.

Vrensen G., De Groot D., and Nuñez-Cardozo J. (1977). Postnatal
development of neurons and synapses in the visual and motor cortex of
rabbits: A quantitative light and electron microscopical study. *Brain Res.
Bull.* 2:405–16.

Weibel E. R. (1979). *Stereological Methods. Vol. 1: Practical Methods for
Biological Morphometry.* Academic Press, London.

  (1980). *Stereological Methods. Vol. 2: Theoretical Foundations.* Academic
Press, London.

White E. L. (1978). Identified neurons in mouse SmI cortex which are
postsynaptic to thalamocortical axon terminals: A combined Golgi-
electron microscopical and degeneration study. *J. Comp. Neurol.*
181:627–62.

  (1989) *Cortical Circuits.* Birkhauser, Boston.

White E. L. and Hersch S. M. (1981). Thalamocortical synapses of pyramidal
cells which project from SmI to MsI cortex in the mouse. *J. Comp.
Neurol.* 198:167–81.

Wolff J. R., Balcar V. J., Zetzsche T., Bottcher H., Schmechel D. E., and
Chromwall B. M. (1984). Development of GABA-ergic system in rat
visual cortex. *Adv. Exp. Med. Biol.* 181:215–36.

Woolsey T. A. and Van der Loos H. (1970). The strucal organization of layer
IV in the somatosensory region (SI) of mouse cerebral cortex. *Brain Res.*
17:205–42.

2

The probability for synaptic contact between neurons in the cortex

This chapter examines techniques for evaluating the probability of finding a synaptic contact between neurons. A simple case for which one might want to estimate that probability is shown in Figure 2.0.1, where neurons from region A send their axons into region B and establish synaptic contacts there.

One of the principal reasons to evaluate the probability of such contact is to compare the evaluated probability and the experimental findings. Such comparisons can reveal the existence of rules that govern the connectivity between neurons. To illustrate this point, let us suppose that we are able to obtain an estimate of the probability of contact between a neuron from A and a neuron from B, assuming complete randomness (i.e., every neuron from A has the same probability of establishing contact with every neuron in B). We can then conduct experiments in which electrodes are thrust into A and B. With one electrode we stimulate a neuron from A, and with the other we record the response of a neuron from B. Such an experiment can help us decide if the cell we stimulate indeed affects the recorded cell synaptically. By repeating the experiment many times, we can experimentally evaluate the probability that a neuron from A will make a functioning synaptic contact on a neuron from B.

If the probability of contact that we determined experimentally agrees with the probability evaluated theoretically, then the idea of completely random connectivity can be accepted. But if they do not agree, there is reason to investigate the source of discrepancy. For instance, such a study may reveal that the connections are governed by a rule that prevents two neurons in A from forming contacts on the same target neuron in B, or it may reveal that there are synapses that appear morphologically normal and yet are not functional. An examination of the causes of discrepancies between the predictions of a model and the experimental

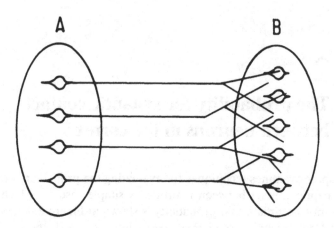

Figure 2.0.1. Neurons from A make synaptic contacts with neurons from B.

results can improve our understanding of connectivity and of the function of neural networks in the brain.

Another purpose of developing a framework for estimating the probability of connections among neurons is to evaluate the probability of the existence of more complex connectivity schemes. For instance, we may wish to evaluate the probability of the connections shown in Figure 2.0.2, in which each of the four presynaptic neurons excites every one of the four postsynaptic neurons.

This chapter proceeds according to the following outline: In Section 2.1 we illustrate how the probability of contact between two neurons can be assessed when only crude measurements of the extent of axonal branching for the presynaptic cell and dendritic branching for the postsynaptic cells are available. The crude estimation method of Section 2.1 is refined in two stages in Sections 2.2 and 2.4. In Section 2.3 we solve several numerical examples, thereby giving the reader some insight into probabilities of contact among neighboring cortical cells.

This chapter may seem tedious to the reader who is not particularly interested in methods for estimating the probability of contact between two neurons. In that case the reader is advised to read Sections 2.1.1 and 2.1.2 and then proceed to Chapter 3.

2.1 Uniform distribution

In this section we develop some crude methods for calculating the probability of contact between two neurons. These methods will be refined in

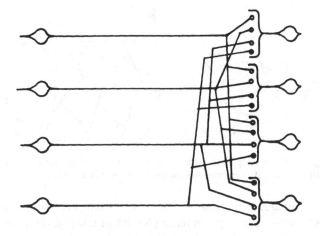

Figure 2.0.2. Every one of the four presynaptic neurons contacts every one of the four postsynaptic neurons.

Sections 2.2 and 2.4. For any given situation, the appropriate level of refinement will depend on the availability of morphological data. Therefore after presenting each computation method, we attempt to describe the available quantitative data and the experimental methods for obtaining them.

2.1.1 Computing the probability of a contact

The basic equation for calculating the probability of contact is obtained by assuming that the presynaptic cell distributes its synapses uniformly within a given volume and by assuming that all the neurons within that volume have the same likelihood of receiving those synapses. Then, if we conduct the following "experiment," selecting at random one neuron from the region and determining whether or not it receives a synaptic contact from the presynaptic cell, we can say that the probability of finding such a neuron is given by

$$\mathrm{pr}\{\text{contact}\} = n/N \qquad (2.1.1)$$

where n represents the number of cells that receive one or more synaptic contacts from the presynaptic axon, and N is the total number of neurons in the region. Most of this section is devoted to techniques for estimating n and N.

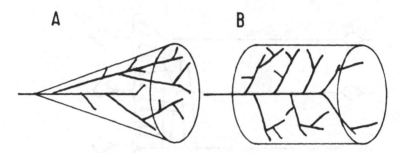

Figure 2.1.1. Examples of axonal ranges. A: The branches are contained within a cone. B: The branches are contained within a cylinder.

In order to evaluate the number of neurons that are candidates for receiving a synapse from a given axon (*N*), let us imagine the axon as it reaches its target zone, where it branches off to form many synapses. Let us look for a simple geometric shape that will include all the branches of the axon. That volume is defined as the axonal range. Figure 2.1.1 illustrates two such axonal ranges. Let V denote the volume of the axonal range, ρ the density of neurons in the region, and N the total number of neurons contained in this volume. Then we have

$$N = \rho V \qquad (2.1.2)$$

Now we need to find the number of cells (*n*) that actually receive synapses from that axon. Under certain favorable conditions, this number can be obtained by direct histological observations, as described in the next example.

Example 2.1.1. Freund et al. [1983] studied chandelier (axo-axonal) cells in great detail. These neurons, which are found mostly in layers II and III, form a subclass of what we call smooth stellate cells (see Chapter 1 for elaboration). In each, the axon ramifies profusely near and under the cell body, as well as at the deeper layers (V and VI). The fine axonal branches terminate in short vertical sections, on which several beads can be seen. Figure 2.1.2B shows part of the axonal range of such a chandelier cell, which was injected with HRP. The fine branches that connect the parent axon to the terminal branches were not drawn, in order not to obscure these terminals.

When the beaded terminals were examined by electron microscopy, they were found to form symmetric (presumably inhibitory) synapses on the initial segments of the pyramidal cells' axons. Each terminal branch of the chandelier cell made one to six synapses with the same pyramidal cell. The chandelier cell shown in Figure 2.1.2B gave synapses to 330 cells in layers II and III, and to an additional 10 cells in the deep layers. Figure 2.1.2C shows the axonal range of this same cell as seen from above. From these two views, we may conclude that the axonal range had the shape of a compressed cone. The base of the cone was an ellipse having axes of 400 and 200 μm, and the cone's height was 300 μm. The volume of this axonal range was about 0.019 mm^3. This cell was in layers II/III of the visual cortex of a cat, with a cell density of about 80,000/mm^3. Most of the cells in these layers are pyramidal. By substituting these values in equation (2.1.2), we discover that the axonal range contained about 1,500 cells, 330 of which were contacted by the chandelier cell. By substituting these values in equation (2.1.1), we get a probability of 0.22 for having a contact between the axon of a chandelier cell and a pyramidal cell (whose cell body is within the axonal range of the chandelier cell).

Example 2.1.1 is a case in which it was relatively easy to determine both the axonal range and the actual number of cells that were contacted by one axon (n). This could be done because each beaded terminal branch of the chandelier cell established contact with one pyramidal cell, so that it was possible to measure n by counting the number of terminal branches of the axon.

In most cases, n must be derived from other observations. It is somewhat easier to obtain an estimate for the total number of synapses that the axon makes (although that also requires much careful work). In Section 2.1.2 we discuss some of the experiments in which such estimates were obtained. For now, let us assume that we find that the axon is involved with m synapses within its axonal range. If $m << N$, and if there is no morphological evidence that one axon makes several repeated synapses with one neuron, we may assume that the axon makes either one synapse per postsynaptic cell or none. In this case, the number of neurons that receive a synaptic contact (n) is equal to the number of synapses (m). Therefore, we may substitute m for n in equation (2.1.1), and the probability of finding a neuron that receives a contact is

$$\text{pr\{contact\}} = m/N \tag{2.1.3}$$

If m is not much smaller than N, we assume that the m synapses are distributed completely at random among the N neurons in the axonal range. On the average, each of these N neurons receives

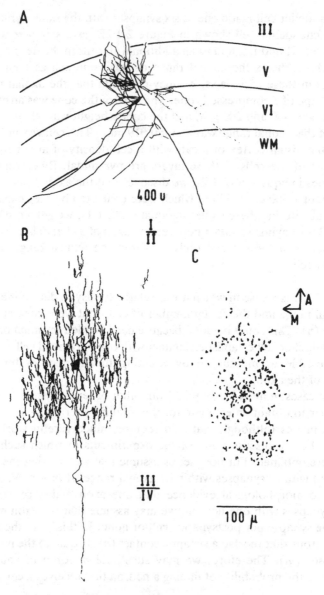

Figure 2.1.2. Distribution of an axonal tree. A: Axonal tree of a pyramidal cell in the cortex. The dendrites of the cell are not drawn. [From P. Landry, A. Labelle, and M. Deschens, "Intracortical Distribution of Axonal Collaterals of Pyramidal Tract Cells in the Cat Motor Cortex," *Brain Research*, 191:327–36, 1980, with permission of Elsevier Science Publishers.]

$$s = m/N \qquad (2.1.4)$$

synapses from the axon. This situation is similar to that of throwing m balls (synapses) into N bins (target neurons). In each bin we expect to find, on the average, s balls (note that s need not be an integer). If we choose one bin at random, we can ask what the probability is of finding it empty, of finding one ball there, of finding two balls, and so forth. These probabilities are given by a Poisson distribution, which tells us the probability of finding k balls when s balls are expected. In our case, if we choose a target neuron at random, what is the probability that it will receive k synapses from the presynaptic axon? This is given by

$$\mathrm{pr}\{\text{receiving } k; \text{ when } s \text{ are expected}\} = \mathrm{pr}\{k; s\} = e^{-s}s^k/k! \qquad (2.1.5)$$

For our purposes, we can deduce the probability of having at least one synaptic contact from the probability of having no contacts ($k = 0$):

$$\mathrm{pr}\{\text{contact}\} = 1 - \mathrm{pr}\{\text{no contact}\} = 1 - \mathrm{pr}\{0; s\} = 1 - e^{-s} \qquad (2.1.6)$$

In many situations the repeated synapses generated by one axon on one postsynaptic neuron are not distributed according to Poisson's law. In experimental studies it is often possible to measure directly the average number of synapses (s) that one axon generates on one postsynaptic neuron. In these situations, if we know how many (total) synapses the axon gives (m), and the average number of repeated synapses it generates on each postsynaptic target (s), we can evaluate the total number of neurons that receive synaptic contacts from the incoming axon (n) as

$$n = m/s \qquad (2.1.7)$$

and the probability of finding a neuron that receives a contact is then given by equation (2.1.1).

Naturally, if the axonal range contains several types of neurons, the probability of contact should be computed separately for each type of cell. For instance, equation (2.1.1) becomes

Caption to Figure 2.1.2 *(cont.)* B: Axonal tree of a chandelier cell in layers II/III of the visual cortex of the cat. Only the terminal branches of the axon are shown, for the sake of clarity. C: View from the pia on the axonal tree of the same axon as in B. [B and C from T. F. Freund, K. A. Martin, A. D. Smith, and P. Somogyi, "Glutamate Decarboxylase Immunoreactive Terminals of Golgi Impregnated Axoaxonic Cells and of Presumed Basket Cells in Synaptic Contact with Pyramidal Neurons of the Cat's Visual Cortex," *Journal of Comparative Neurology,* 221:263–78, 1983, with permission of Alan R. Liss, Inc.]

$$\text{pr}_i = n_i/N_i, \quad i = 1, 2, \ldots, l$$

where pr_i denotes the probability that the axon contacts a neuron of type i, n_i is the number of neurons of type i on which the axon synapses, N_i is the total number of neurons of type i in the axonal range, and l is the number of different types of neurons in the axonal range. Equation (2.1.2) becomes

$$N_i = \rho_i V, \quad i = 1, 2, \ldots, l$$

and so on.

The equations derived thus far present a rather poor approximation of the real situation in the cortex, because they ignore two important factors: The synapses generated by an axon usually are not limited to neurons whose cell bodies are within the axonal range, because neurons located at the fringe of the axonal range usually have dendrites that extend into the axonal range and may also receive synapses. Furthermore, the density of the synapses within the axonal range need not be uniform. As we proceed with this chapter, the equations will be refined to include these factors. But before doing so, we examine in Section 2.1.2 some of the measurement techniques available for the parameters described thus far.

2.1.2 Experimental estimation of the number of postsynaptic neurons contacted by one presynaptic axon

In this section we discuss experimental ways to determine the axonal range (V), the total number of cells that receive at least one synapse from the axon (n), the total number of synapses generated by the axon (m), and the average number of repeated synapses between the single axon and one target cell or its dendrites (s).

2.1.2.1 Axonal range. Determining the axonal range is fairly easy. It is accomplished by staining the axon and all its branches and estimating the volume in which they are distributed by examining thick sections (100 μm or more). In the past, most estimates of axonal ranges were obtained by observing neurons stained by Golgi impregnation. But that method cannot assure growth of the stain crystal into the entire extent of the finer branches of the axon. When the axon becomes myelinated, the stain does not impregnate it. Despite many improvements in the Golgi method introduced over the years, one should regard Golgi measurements of axonal range (and length) as underestimates of the true axonal extent. See Section 1.3.1 for a more detailed discussion of this issue.

In recent years, several investigators have succeeded in injecting dyes into single neurons via a micropipette. Some of these substances are carried into the axon and all its branches by the axonal anterograde transport. This is the mechanism by which the cell body provides substances to the axon. If one waits long enough after the injection of such a dye, one can expect to see it in the full extent of all the branches of the axon. Figure 2.1.2 shows two examples of such experiments.

2.1.2.2 Candidate target cells (N) and actual target cells (n). Once we have evaluated the volume (V) of the axonal range, we need to know the density of neurons (ρ) in order to compute the number (N) of candidate targets, according to equation (2.1.2). These densities can be determined from examination of histological sections, as described in Section 1.4. To complete our calculation of the probability of contact, we need the number of target neurons (n) on which the axon actually generates synaptic contacts. However, direct estimation of this number is extremely tedious. Only in special situations can it be carried out directly, as described in Example 2.1.1.

2.1.2.3 Experimental determination of m. Direct determination of the total number of synapses that an axon generates (m) is difficult. Most of the axons in the cortex generate synapses along their unmyelinated portions (not just at the end of a branch). The location of such an "en passant" synapse usually is marked by a slight varicosity of the axon. One must remember, though, that only after the Golgi precipitate is replaced by a more delicate stain (usually gold) and the region is reexamined by electron microscopy [Fairen, Peters, and Saldanha, 1977] can one be sure that it is a real synapse. One should also bear in mind that at one varicosity, an axon may establish synaptic contacts with several distinct postsynaptic elements [Somogyi et al., 1983].

The total number of synapses that an axon generates (m) can be estimated using a method suggested by Braitenberg: The total length of the axon and its branches is measured (e.g., in HRP or similar preparations), and the average distance between two successive synapses on the axon is estimated using electron microscopy. The total number of synapses per axon can be computed from these two figures.

Example 2.1.2. According to Braitenberg [1978a], the total length of the local branches of a pyramidal cell in the cortex of the mouse is about 10 mm. The average distance between two adjacent synapses is about 4 μm. Therefore, each pyramidal cell in the mouse generates 2,500 synapses in its vicinity.

2.1.2.4 Experimental determination of s. The number of repeated synapses on a given neuron [s in equations (2.1.4) and (2.1.5)] can be determined experimentally only in favorable cases. The synapses generated by the chandelier cell on the pyramidal cell, as described in Example 2.1.1, present such a favorable case. In that case, all the synapses generated on a given target cell were concentrated in one small region. The multipolar smooth stellate cell (also called basket cell) presents a more complicated case, as described next.

Example 2.1.3. Somogyi et al. [1983] examined in detail the axons of three basket cells from layers III and IV of the visual cortex of the cat. HRP was injected into these cells in the course of an electrophysiological study of their response properties. In each case the axon branched out extensively in the vicinity of the cell body and formed many synapses throughout its range. The synapses were found both at varicosities along the axon and at the endings of clublike short processes. All the synapses that were examined by electron microscopy were of the symmetric type (presumably inhibitory synapses). Of the 177 postsynaptic elements that were analyzed, 64 (36 percent) were on the somata of neurons, and 27 (15 percent) terminated on dendritic shafts that were close to the cell body. The remainder of the synapses terminated on more remote dendrites.

Thus, aproximately half of the synapses of these basket cells were localized on small targets: the cell bodies and proximal dendrites. Most of the target cells were pryamidal neurons. On the average, each pyramidal neuron received 4.4 synapses from a given basket cell on its cell body and an additional 2.3 synapses on its proximal dendritic shafts. Thus, for the synaptic contacts between a large basket cell and the pyramidal cell body (and vicinity), the average number of repeated synapses was about 7.

If we wish to consider synapses on distal dendrites, we often need to resort to indirect estimations, such as suggested by equation (2.1.4). This equation assumes that the synapses are randomly distributed among all (candidate) targets, and it estimates the expected number of repeated synapses by dividing the total number of synapses generated by an axon (m) by the total number of target cells in its range (N).

This estimate of the number of repeated synapses may be biased in two directions. If an internal mechanism exists that prohibits the presynaptic neuron from generating additional synapses on a given target once it has generated a synapse, then s as computed by equation (2.1.4) is an overestimate of the number of repeated synapses. If, on the other hand, a mechanism exists that attracts additional synapses to a given target cell once a contact is established (as is probably the case for the chandelier cell in Example 2.1.1), then s is an underestimate. Most of the relevant anatomical observations, however, suggest that s is an underestimate of the expected number of repeated synapses.

If we wish to use the total number of synapses of an axon to determine the total number of cells that receive synaptic contacts from that axon [equation (2.1.4)], we have to estimate the number of repeated synapses generated by a given axon on a given target cell (s). In the case of the basket cell described in Example 2.1.3, it was possible to state the number of repeated synapses between the axon and a given target neuron as long as the target area was limited to the cell body or the proximal dendrites. But approximately half of the synapses generated by the basket cell were located on distal dendrites, where it is difficult to distinguish which of the dendrites belong to the given parent cell body. One must bear in mind that more than 99 percent of cortical synapses are found on dendrites. Thus, excluding the few special cases in which the target is selectively the cell body or its vicinity, it is difficult to observe directly the number of repeated synapses between one axon and one target cell. In the examples that follow, several such estimates are given.

Example 2.1.4. Marin-Padilla [1968] described contacts between axons of unidentified origin and pyramidal cells in the cortex of a two-month-old human baby. He found that a given axon tended to generate many repeated synapses with the apical dendrite of a pyramidal cell, but far fewer on the basal dendrite. In his material, an axon generated twenty to sixty (thirty-five on the average) synapses with the apical dendrite, but only five to ten with the basal dendrite. One must bear in mind that the connectivity in a two-month-old baby does not represent that of the mature cortex.

Example 2.1.5. Braitenberg [1978b] computed the expected number of synapses between the axon of a pyramidal cell and the basal dendrites of a nearby pyramidal cell. Using morphometric parameters derived from mice, he concluded that the probability of obtaining more than one synapse between an axon and the dendrites of a nearby cell was between 0.0006 and 0.27, depending on the cell's position.

The large discrepancy between the observations of Example 2.1.4 and the statistical estimate of Example 2.1.5 may be attributed to two sources. Marin-Padilla described the case of an axon that runs parallel to the apical dendrites of cortical pyramidal cells. When such an axon happens to be in close apposition to an apical dendrite, it has a high probability of establishing multiple synapses with that dendrite. Braitenberg's calculations apply to local, intracortical axons of pyramidal cells, which tend to take a straight course that is not oriented in any preferred direction. In addition, one must note that Braitenberg attempted to calculate average probabilities, whereas Marin-Padilla described particularly extravagent cases. Martin [1988] suggested that only basket cells tend to make multiple repeated (inhibitory) synapses on the apical dendrites of pyramidal cells.

Example 2.1.6. Perhaps the most direct observations of the synaptic contact between an afferent axon and the dendrites of its target cells have been made in the spinal cord. Several researchers injected HRP into a single afferent coming from a muscle spindle (called Ia afferent) and into some of its target cells, motoneurons that innervate either the same muscle or synergic muscles. After such injections it was possible to identify points of intimate contact between any branch of a stained axon and any dendrite of a given motoneuron. Brown and Fyffe [1981] found that a single Ia afferent axon had, on the average, 2 contact points with the cell body and 3.4 contact points with dendrites. They considered contacts as loci where the axon had a varicosity or a terminal bouton. Burke et al. [1980] reported a median of 7 contacts between a Ia afferent axon and a homonymous motoneuron, and 4 contacts between a Ia afferent axon and a heteronymous motoneuron. Such observations are, unfortunately, not yet available for the cerebral cortex.

Most of the cortical synapses are excitatory, with the overriding majority being contributed by cortical pyramidal cells and spiny stellate cells (either locally or from other cortical regions). To date, there have been no direct observations regarding the mode of distribution of these synapses. In the absence of evidence to the contrary, we assume that these

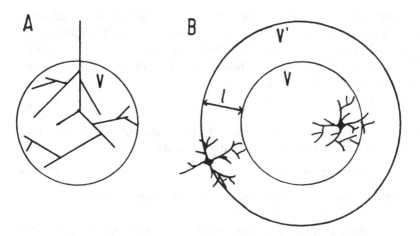

Figure 2.1.3. Correcting the range of an axon by considering the size of the dendritic tree of the receiving cells. A: The axonal range. B: The extended range.

synapses are formed according to certain simple probabilistic laws; yet we must bear in mind that not all the synapses are formed in this way. The chandelier cell synapses (Example 2.1.1) and the basket cell synapses (Example 2.1.2) are two notable exceptions. It is worth noting that these two documented exceptions are (presumably) inhibitory synapses.

The remainder of this chapter consists of an elaboration of the simple equations derived in Section 2.1.1 and includes some examples of their use. The reader who is not specially interested in computing the probability of contact may proceed directly to Chapter 3.

2.1.3 Correction for the extension of the dendritic tree

Let us now return to the computation of the probability of contact between an axon and neurons. Earlier we described how the axon arrives at its target zone and branches out within a volume that we call the axonal range. We assume that the axon may generate synapses on neurons whose cell bodies are located inside this axonal range. This assumption is accurate in certain special cases, such as the case of the chandelier cell described in Example 2.1.1, but in most cases the axon can also establish synaptic contacts with dendrites that extend into the axonal range, dendrites whose cell bodies are outside the axonal range. This situation is illustrated in Figure 2.1.3.

If the dendrites extend, on the average, a distance *l* from the cell body, then in order to obtain the range of potential contact of the axon, one must add a margin whose width is *l* to the axonal range. If we denote by *V'* the volume of this extended range, then the number of cells that may be contacted by the axon is

$$N = V'\rho \qquad (2.1.8)$$

where ρ denotes the density of the cells in *V'*. Accordingly, the probability of contact between the axon and a given cell in *V'* is

$$\text{pr}\{\text{contact}\} = n/(V'\rho) \qquad (2.1.9)$$

where *n* is the total number of cells with which the axon synapses.

Example 2.1.7. The local collaterals of an axon of a pyramidal cell in the cortex of the cat extend, on the average, a distance of 0.25 mm on each side of the pyramidal cell body. Let us assume that the local axonal range of the pyramidal cell is a sphere with a radius of 0.25 mm. The dendrites of the nearby cells also extend to a distance of 0.25 mm from their cell bodies. Thus, the extended axonal range is a sphere with a radius of 0.5 mm. This extended range has a volume of about 0.5 mm^3 and contains about 20,000 neurons.

The total local length of the axon of a pyramidal cell in the cat is about 36 mm, and it has a synapse about ever 3 μm. It therefore generates approximately 12,000 synapses. Let us assume that the multiplicity of contacts to a given cell is two to three; then this axon contacts approximately 5,000 cells in its vicinity. Thus, the probability that a cell whose cell body is in the extended axonal range will receive a synapse from that axon is 0.25 (5,000/20,000).

In an experiment performed by Abeles [1982], adjacent neurons in the auditory cortex of the cat were recorded. Evidence of direct synaptic contact was found for 12 percent of the studied pairs, by constructing cross-correlations between cells. This means that if we select two adjacent cells (*a* and *b*) and inquire as to the probability that either *a* is presynaptic (to *b*) or *b* is presynaptic (to *a*), we arrive at 0.12. The anatomical estimate says that in this situation the probability of noting synaptic contact is $(0.25 + 0.25 - 0.25 \cdot 0.25) = 0.44$. The experimentally derived probability of contact is considerably smaller than the morphological estimation of 0.44. The discrepancy is even larger if we consider that the probability of contact between adjacent neurons ought to be higher than the average probability for the entire extended axonal range, and that the multiplicity of contacts between excitatory cortical neurons is likely to be less than two to three.

The calculations of the probability of contact in equation (2.1.9) and in Example 2.1.7 are inappropriate because they assume uniform density of contacts within the extended axonal range. That is, we assumed that all the neurons whose cell bodies were within the extended range had equal probabilities of receiving a synapse, and that the axonal branches were distributed at uniform density in the axonal range. In the next section, we refine the equations to consider inhomogeneities.

2.2 Nonuniform distribution

In this section we relax the constraint that all the neurons within the extended range of an axon must have the same probability of receiving a synapse from that axon. To do so, we first assess the expected number of synapses between a given axon and the dendritic tree of a given neuron. Then, using equation (2.1.6), we compute the probability of contact between the axon and the neuron.

Let us choose a certain axon as a presynaptic source, and a certain neuron, whose dendrites extend into the axonal range of that axon, as a candidate postsynaptic neuron. It would be impossible to identify all the synapses between every possible pair of axon and neuron that we might choose. But we can make reasonable estimates of the density of synapses generated by the presynaptic axon (on all sorts of cells) and the density of synapses received by the postsynaptic neuron (from all sorts of sources). These densities can be used to compute the expected number of synapses between the axon and the neuron.

Let us first look at a very small cube of tissue measuring $dx \cdot dy \cdot dz$ that is positioned at (x, y, z). For brevity, we denote this volume element by dX, and its position by X. Within this small cube we have all sorts of synapses, some of which belong to our presynaptic axon and some of which belong to the dendrites of the postsynaptic neuron.

Let us now randomly select one synapse from within dX and inquire as to the probability that this synapse belongs to the dendrites of the postsynaptic neuron. If the element of volume dX is small enough, we can expect that the probability that the synapse picked will belong to our postsynaptic neuron will not vary through dX. Therefore, the probability that a synapse from within dX will be on the dendritic tree of our neuron is

$$\text{pr\{dendritic\}} = dm/dN \qquad (2.2.1)$$

where dm denotes the expected number of synapses on the dendritic tree of our cell in dX, and dN is the total number of synapses in dX. These two numbers are given by

$$dN = \rho \, dX \qquad (2.2.2)$$

and

$$dm = \mu(X) \, dX \qquad (2.2.3)$$

where ρ denotes the density of synapses in the cortex, and $\mu(X)$ is the density of synapses on the dendritic tree of our neuron. Note that we assume here that the total synaptic density (ρ) is constant, whereas the density of synapses received by the cell [$\mu(X)$] may change as a function of position (depending on the pattern of branching of the dendrites).

Similarly, if $\nu(X)$ denotes the density of the synapses that are generated by the axon, dn the expected number of synapses that will be generated by the axon and be located inside dX, and pr{axo} the probability that given synapse within dX will come from our axon, we have

$$dn = \nu(X) \, dX \qquad (2.2.4)$$

and

$$\text{pr\{axo\}} = dn/dN \qquad (2.2.5)$$

We assume that the event {axo} = {we randomly selected a synapse within dX and it belonged to our axon} and the event {dendritic} = {we randomly selected a synapse from within dX and it belonged to the dendrites of our neuron} are independent. Then we have the probability of the event {axodendritic} = {we picked at random a synapse from within dX and it was a synapse between our axon and a dendrite of our neuron} given by

$$\text{pr\{axodendritic\}} = \text{pr\{axo\}} \cdot \text{pr\{dendritic\}} = dm \, dn/dN^2 \quad (2.2.6)$$

From this probability we can compute the expected number of synapses between our axon and the dendrites of our neuron that are within dX:

$$ds = \text{pr\{axodendritic\}} \, dN = dn \, dm/dN = [\mu(X)\nu(X)/\rho]dX \qquad (2.2.7)$$

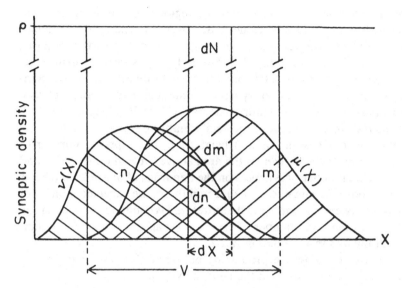

Figure 2.2.1. Synaptic densities in a unidimensional example.

We define the dendritic domain of a neuron in similar fashion as the axonal range. That is, it is a region with a simple geometric shape that contains all the dendritic branches of the neuron. Then, if V denotes the intersection of the axonal range of our axon with the dendritic domain of our neuron, and s is the expected number of synapses between our axon and neuron, we have

$$s = \int_V ds = (1/\rho) \int_V \mu(X)\nu(X) \, dX \tag{2.2.8}$$

In addition, we have two limiting conditions:

$$m = \int_{dd} \mu(X) \, dX \tag{2.2.9}$$

$$n = \int_{ar} \nu(X) \, dX \tag{2.2.10}$$

where m denotes the total number of synapses made on the dendrites of the postsynaptic neuron, dd is the dendritic domain of that neuron, n is the total number of synapses that our axon generates, and ar is its axonal range.

Figure 2.2.1 illustrates these formulas for unidimensional distributions. The presynaptic axon arrives at a region where it distributes synapses with varying density, as described by $\nu(X)$. The total area under this curve gives the total number of synapses generated by the axon (n), as described by equation (2.2.10). Similarly, a postsynaptic neuron receives synapses from all kinds of sources, with density $\mu(X)$. We compute the expected number of synapses in the region of overlap between the axonal range and the dendritic domain (V) by breaking it into small sections dX. In each section, the total number of synapses (dN) is the rectangle whose base is dX and height is ρ; the expected number of synapses from the axon (dn) is the area above dX and under $\nu(X)$, and the expected number of synapses on the dendrites of our neuron (dm) is the area above dX and under $\mu(X)$. From these, the expected number of synapses between the axon and dendrites (ds) is computed with equation (2.2.7), and the total number of expected synapses is obtained by summing all the ds's in V [equation (2.2.8)].

If we wish to know the probability of contact (by one or more synapses) between the axon and the neuron, we use

$$\text{pr\{contact\}} = 1 - \text{pr\{no contact\}} = 1 - e^{-s}$$

as described in Section 2.1.

2.3 Examples

In this section we illustrate how the equations developed in the two preceding sections might be used to evaluate intracortical connectivity.

2.3.1 Uniform synaptic density

Example 2.3.1. Given that (1) the axonal range of a certain presynaptic axon is a sphere whose radius is 0.25 mm, and within that range the neuron generates 12,000 synapses that are uniformly distributed throughout its range [i.e., $\nu(X) = $ constant], (2) the dendritic domain of the basal dendrites of a certain postsynaptic cortical cell is also a sphere with a radius of 0.25 mm, within which the neuron receives 20,000 synapses that are distributed uniformly throughout the domain [i.e., $\mu(X) = $ constant], (3) the total synaptic density within the cortex is $0.8 \cdot 10^9$ synapses per cubic millimeter, and (4) the cell body of the postsynaptic neuron is at the center of the axonal range of the presynaptic axon, then what is the expected number of synapses between the presynaptic axon and the basal dendritic tree of the postsynaptic neuron?

Solution. The volume of the axonal range (and of the dendritic domain) is about 0.06 mm^3; therefore, the synaptic densities of the axon and of the dendrites are

$$\nu(X) = \nu = 12{,}000/0.06 = 200{,}000 \text{ synapses/mm}^3$$
$$\mu(X) = \mu = 20{,}000/0.06 = 333{,}000 \text{ synapses/mm}^3$$

Because the densities are constant, equation (2.2.8) is reduced to

$$s = (\mu\nu/\rho) \int_V dX = \mu\nu V/\rho = 200{,}000 \cdot 333{,}000 \cdot 0.06/8 \cdot 10^8 = 5$$

Thus, we can expect to find by chance about five synapses between an axon and a neuron that is optimally placed within the axonal range.

Example 2.3.2. In electrophysiological experiments [Abeles, 1982], the probability of finding synaptic interaction between two adjacent neurons was 0.12. Is this result compatible with our estimate of five synapses, on the average, between one presynaptic axon and the basal dendrites of a postsynaptic neuron?

Solution. The electrophysiological results mean that the probability of not finding a synaptic contact between the two cells is 0.88 (1 − 0.12). If, before measurement, we should try to guess which neuron would be the presynaptic, and which the postsynaptic, we would arrive at a probability of only 0.06 for guessing correctly, but 0.94 for guessing incorrectly.

If we assume that when the two cell bodies are adjacent, the axonal range of one of the neurons overlaps the basal dendritic domain of the other neuron, then we can compare (a) the probability of no contact (0.94) between the axon of one of the cells and the other neuron and (b) the estimated number of synapses (five) between that axon and the basal dendrites of the other neuron. To do so, we refer once more to the Poisson formula, which states the probability of obtaining m events when the expected number of events is s:

$$\text{pr}\{m; s\} = e^{-s}s^m/m!$$

In our case, the probability of having no synaptic contact when the expected number is five will be

$$\text{pr}\{0; 5\} = e^{-5} < 0.01$$

The discrepancy between 0.01 and 0.94 cannot be accounted for by simple experimental error.

Exercise 2.3.1. In the electrophysiological experiments described in Example 2.3.2, the existence of a synapse was determined by cross-correlating the spike trains of two adjacent neurons. Note that the cross-correlation technique is not sensitive to weak interactions. Assuming that a synaptic relation could be detected in the foregoing experiments only if the presynaptic axon made m or more synapses on a given postsynaptic neuron, what is the smallest m for which the experimental results are compatible with the expected number of synapses (i.e., five)?

Example 2.3.3. According to Braitenberg [1978a], the axonal range of the local axons of a pyramidal cell is below the cell body, whereas the dendritic domain of the basal dendrites of a pyramidal cell is centered around the cell body. This could be the source of the discrepancy between the anatomical and electrophysiological findings described in Example 2.3.2. What degree of overlap between the axonal range and the dendritic domain would be compatible with the electrophysiological results?

To answer this question, we must first evaluate the expected number of synapses compatible with a probability of no contact of 0.94:

$$\text{pr}\{0; s\} = e^{-s} = 0.94$$
$$s = -\ln(0.94) = 0.062$$

This means that if only 0.062 synapse were expected, we would expect to find a probability of no contact of 0.94. Now we can determine from equation (2.2.8) the volume of overlap V between the axonal range and the dendritic domain that would account for an expected number of synapses of 0.062:

$$0.062 = \mu\nu V/\rho = 200{,}000 \cdot 333{,}000V/8 \cdot 10^8 = 83V$$
$$V = 0.00075 \text{ mm}^3$$

The volume of the axonal range (or the dendritic domain) was 0.06 mm³; thus, the overlapping part is only 1.3 percent (0.00075/0.06) of the axonal range. This means that if we can detect even one synaptic contact by cross-correlation, the axonal ranges and dendritic domains of adjacent neurons must be almost completely disjointed.

2.3.2 Nonuniform synaptic densities

Sholl [1956] evaluated the densities of basal dendrites in the following manner: He stained the visual cortex of a cat with the Golgi method. For several neurons, he measured the length of each dendritic branch and its points of bifurcation. From that information, assuming that the dendrites were distributed in a radial fashion, he could calculate how many branches he would expect to find at distances of 50 μm, 100 μm, and so forth, from the cell body. Imagine a set of spheres centered about the cell body. Each such sphere has a surface area of $4\pi r^2$, where r is the distance from the cell body. By dividing the number of dendritic branches at a distance r from the cell body by the surface area of a sphere of radius r, he computed the density of branches at a distance r from the cell body. His data showed that the density falls exponentially, with a spread constant of about 80 μm.

If we make the (reasonable) assumption that the density of synapses along a dendritic branch is constant (see Figures 3.6 and 3.7 of White [1989] for a refinement of this assumption), then it follows that the density of synapses on the dendritic tree behaves as follows:

$$\mu(r) = ae^{-kr} \tag{2.3.1}$$

The proportionality constant (a) can be evaluated from equation (2.2.9), which after being converted to radial coordinates reads

$$m = \int_0^x 4\pi r^2 \mu(r)\, dr = 4\pi a \int_0^x r^2 e^{-kr} dr = 8\pi a/k^3$$

We can solve this for a:

$$a = mk^3/(8\pi) \tag{2.3.2}$$

Example 2.3.4. The total number of synapses on the basal dendritic tree is 20,000, and the spread constant (k^{-1}) is 0.08 mm. What is the proportionality constant (a) of equation (2.3.1)?

By using equation (2.3.2) we arrive at

$$a_m = 20{,}000/(8\pi 0.08^3) = 1.6 \cdot 10^6 \text{ synapses/mm}^3$$

Therefore, $\mu(r)$ is given by $\mu(r) = 1.6 \cdot 10^6 \cdot e^{-r/0.08}$.

If we assume that the local branches of an axon are distributed in a similar manner, but the axon makes only 12,000 synapses, we arrive at

$$a_n = 12{,}000/(8\pi 0.08^3) = 1 \cdot 10^6 \text{ synapses/mm}^3$$

Therefore, $\nu(r)$ is given by

$$\nu(r) = 10^6 \cdot e^{-r/0.08}$$

Example 2.3.5. There is total overlap between an axonal range and a dendritic domain, with the same parameters as described in Example 2.3.4. What is the expected number of synapses between the axon and the dendrites?

For the case of radial symmetry, equation (2.2.7) becomes

$$ds = 4\pi r^2 \mu(r) 4\pi r^2 \nu(r)/(4\pi r^2 \rho) \, dr \qquad (2.3.3)$$

where ds now represents the expected number of synapses between the axon and the dendrites within a thin spheric shell whose width is dr and whose distance from the center is r. The total expected number of synapses is given by

$$s = \int_0^\infty ds = 4\pi a_n a_m/\rho \int_0^\infty r^2 e^{-2kr} dr = \pi a_n a_m/(\rho k^3) = 3.2 \text{ synapses} \quad (2.3.4)$$

Note that when uniform densities were assumed in Example 2.3.1, we arrived at an expected number of five synapses between the axon and the dendrites. The two estimates are not too distant from one another. Both models give estimates close to 1 for the probability of contact between the axon and a neuron (which are optimally placed). This is much higher than the probability of 0.25 estimated in Example 2.1.7, in which we computed the average probability of contact between an axon and a neuron situated anywhere in the axon's extended range.

Exercise 2.3.2. Assuming that the densities of synapses are as described in Example 2.3.4, but the dendrites are not optimally located for receiving a synapse, what distance between the center of the axonal range and the center of the dendritic domain will give a probability of contact between the axon and the dendrites of 0.06?

Exercise 2.3.3. Assuming that the density of the dentritic synapses is as described in Example 2.3.4, but the axonal range is not a radially symmetric sphere, but a cylinder whose height is 0.5 mm and whose radius is 0.25 mm, and assuming also that within this cylinder the density of the axonal synapses is constant, what will be the average number of synapses between this axon and the dendrites of a cell whose cell body is in the cylinder's center?

In this section, simple symmetric distributions of branches have been assumed. Krone et al. [1986] described more realistic dendritic domains and axonal ranges; they also used symmetric distributions. However, that is not always the case. Glaser, Van der Loos, and Gissler [1979] measured the distribution of dendritic trees in the auditory cortex in cats. They found that the distribution was slightly biased in one direction. They called their dendritic domains "trumpets." Unfortunately, they did not give a quantitative description of how the dendritic density varied within the trumpet.

2.4 Generalization

In this section we describe two ways to generalize the equations for computing the expected number of synapses between an axon and the dendrites of another cell. In the first case, we allow the background synaptic density to vary, and in the second case we consider the possibility of different affinities between different types of cortical neurons.

2.4.1 Nonuniform background density

In Section 2.2 we allowed the density of the synapses of the axon [$\nu(X)$] and the density of the synapses on the dendrites [$\mu(X)$] to vary as a function of location, but considered the density of all the synapses in the region (ρ) constant. However, when the axonal range (or the dendritic domain) extends through more than one cortical layer, the density of the synapses is likely to vary. In order to take this into consideration, we should allow ρ to vary as well.

Let X denote the location in the brain, and dX a small element of volume around it. Then equation (2.2.7) will become

$$ds = \mu(X)\nu(X) \, dX/\rho(X) \tag{2.4.1}$$

where ds is the expected number of synapses between the presynaptic axon and the dendrites of the postsynaptic neuron that are within the small cube dX. By integrating equation (2.4.1) we get the total expected number of synapses s:

$$s = \int_V [\mu(X)\nu(X)/\rho(X)] \, dX \qquad (2.4.2)$$

where V again represents the volume of overlap between the axonal range and the dendritic domain.

We must bear in mind that changes in total synaptic density among the six cortical layers are small [Vrensen, De Groot, and Nuñes-Cardozo, 1977], and therefore there is no great advantage in allowing ρ to vary as a function of location. However, if we wish to deal only with a certain type of synapse (as described later), then ρ should indeed be allowed to vary.

2.4.2 Connectivity between specific types

In general, one cannot expect the axon to be "indifferent" as to which cell it contacts. A smooth stellate cell may prefer, for instance, to establish synapses with pyramidal cells. In the following section, such possible specificities are taken into account. The various cell types are denoted by the indices i and j, such that i and j can assume various values (e.g., a pyramidal cell, a smooth stellate cell, a spiny stellate cell, a double-bouquet cell).

Let us assume that we have an axon of type j arriving at a region, branching off, and establishing synapses with the dendrites of a cell of type i. What is the expected number of synapses (s_{ij}) between one particular axon of type j and the dendrites of a given neuron of type i? To answer this question, we need to know the following parameters:

1. $\nu_{ij}(X)$ – the density of synapses from the particular axon (of type j), whose postsynaptic components are dendrites of neurons of type i
2. $\mu_{ij}(X)$ – the density of synapses found on the dendrites of the given neuron (of type i), whose presynaptic components are axons of type j
3. $\rho_{ij}(X)$ – the total density of synapses between all the axons of type j in the region and all the dendrites of type i in the region

Then equation (2.4.2) becomes

$$s_{ij} = \int_V [\mu_{ij}(X)\nu_{ij}(X)/\rho_{ij}(X)] \, dX \qquad (2.4.3)$$

It should be noted that it is possible to arrive at reasonable estimates of all these densities by quantitatively examining brain sections. For example, White and Rock [1981], White and Hersch [1982], and White, Benshalom, and Hersch [1984] were able to determine that within layer IV of the primary somatosensory cortex of the mouse, 18 percent of the synaptic profiles came from specific thalamic afferent axons. They also found that within layer IV, 17 percent of the synapses found on the dendrites of bitufted stellate cells involved these thalamic afferents, 16 percent of the synapses found on the dendrites of spiny stellate cells involved these thalamic afferents, and 13 percent of the synapses found on the dendrites of corticothalamic pyramidal cells involved these thalamic afferents, and 4 percent of the synapses found on the dendrites of corticocortical pyramidal cells involved these thalamic afferents. See White [1989] for a summary of these findings and their significance. Thus, it is possible to determine the densities required for solving equation (2.4.3) by combining such measurements with quantitative data on the geometry of dendrites and cell type densities in layer IV. Unfortunately, quantitative studies of this sort have been rare.

Exercise 2.4.1. The following parameters are given:

1. Thalamocortical afferent axons arrive at layer IV from below; 10,000 such axons arrive through every square millimeter of tangential cortical area.
2. Each thalamic afferent branches out profusely within layer IV. Its axonal range has the form of a cylinder whose height occupies the entire thickness of layer IV (350 μm); the base of this cylinder has a diameter of 300 μm [Landry, Villemure, and Deschens, 1982]. The synapses generated by the thalamocortical afferents are distributed uniformly within this volume.
3. The total synaptic density in layer IV is $8 \cdot 10^8$/mm^3; 18 percent of these synapses are generated by the thalamic afferents.
4. The dendrites of a spiny stellate cell in layer IV are distributed with an exponentially falling density, as described by Sholl [equation (2.3.1)], with a spread constant of 0.06 mm. The total number of synapses that the cell receives on its dendrites is 15,000; 16 percent of these come from the thalamic afferents.
5. The density of the spiny stellate cells in layer IV is 20,000 per cubic millimeter.

From these data, compute the densities ρ_{ij}, μ_{ij}, and ν_{ij}, where j refers to a thalamic afferent axon and i represents a spiny stellate cell. What is the expected number of repeated synapses generated by one thalamic afferent on one spiny stellate cell whose cell body is at the center of the axonal range of the thalamic afferent?

Most of the equations and calculations presented in this chapter are based on the assumption that synapses are distributed in a Poisson-like manner. That is to say, the fact that a certain axon generates a synapse on a dendrite of a certain neuron does not affect its probability of generating another synapse on the dendrites of the same neuron. This is certainly not the case for the axon of a chandelier cell or a basket cell (as described in Section 2.2). However, we must note that most of the cortical synapses are generated by axons of pyramidal cells, in which case it appears that the Poison approximation is reasonable.

2.5 References

Abeles M. (1982). *Studies of Brain Function. Vol. 6: Local Cortical Circuits: An Electrophysiological Study*, pp. 27–32. Springer-Verlag, Berlin.

Braitenberg V. (1978a). Cortical architectonics: General and areal. In Brazier M. A. B. and Petche M. (eds.), *Architectonics of the Cerebral Cortex*, pp. 443–65. Raven Press, New York.

  (1978b). Cell assemblies in the cerebral cortex. In Heim R. and Palm G. (eds.), *Lecture Notes in Biomathematics, Vol. 21: Theoretical Approaches to Complex Systems*, pp. 171–88. Springer-Verlag, Berlin.

Brown A. G. and Fyffe R. E. W. (1981). Direct observations on the contacts made between Ia afferent fibers and α-motoneurons in the cat's lumbosacral spinal cord. *J. Physiol. (Lond.)* 313:121–40.

Burke R. E., Pinter M. J., Lev-Tov A., and O'Donovan M. J. (1980). Anatomy of monosynaptic contacts from Ia afferents to defined types of extensor α motoneurons in the cat. *Neurosci. Abst.* 6:713.

Fairen A., Peters A., and Saldanha Y. (1977). A new procedure for examining Golgi impregnated neurons by light and electron microscopy. *J. Neurocytol.* 6:311–37.

Freund T. F., Martin K. A., Smith A. D., and Somogyi P. (1983). Glutamate decarboxylase immunoreactive terminals of Golgi impregnated axoaxonic cells and of presumed basket cells in synaptic contact with pyramidal neurons of the cat's visual cortex. *J. Comp. neurol.* 221:263–78.

Glaser E. M., Van der Loos H., and Gissler M. (1979). Tangential orientation and spatial order in dendrites of cat auditory cortex: A computer microscope study of Golgi impregnated material. *Exp. Brain Res.* 36:411–31.

Krone G., Mallot H., Palm G., and Schuz A. (1986). Spatiotemporal receptive fields: A dynamical model derived from cortical architectonics. *Proc. R. Soc. Lond. [Biol.]* 226:421–44.

Landry P., Labelle A., and Deschens M. (1980). Intracortical distribution of axonal collaterals of pyramidal tract cells in the cat motor cortex. *Brain Res.* 191:327–36.

Landry P., Villemure J., and Deschens M. (1982). Geometry and orientation of thalamocortical arborization in cat somatosensory cortex as revealed by computer reconstruction. *Brain Res.* 237:222–6.

Marin-Padilla M. (1968). Cortical axo-spino dendritic synapses in man: A Golgi study. *Brain Res.* 8:196–200.

Martin K. A. C. (1988). From single cells to simple circuits in the cerebral cortex. *Q. J. Exp. Physiol.* 73:637–702.

Sholl D. A. (1956). *The Organization of the Cerebral Cortex*. Methuen, London.

Somogyi P., Kisvarday Z. F., Martin K. A. C., and Whitteridge D. (1983). Synaptic connections of morphologically identified and physiologically characterized large basket cells in the striate cortex of cat. *Neuroscience* 10:261–94.

Vrensen G., De Groot D., and Nuñes-Cardozo J. (1977). Postnatal development of neurons and synapses in the visual and motor cortex of rabbits: A quantitative light and electron microscopic study. *Brain Res. Bull.* 2:405–16.

White E. L. (1989). *Cortical Circuits: Synaptic Organization of the Cerebral Cortex. Structure, Function and Theory*. Birkhauser, Boston.

White E. L., Benshalom G., and Hersch S. M. (1984). Thalamocortical and other synapses involving nonspiny multipolar cells of mouse smI cortex. *J. Comp. Neurol.* 229:311–20.

White E. L. and Hersch S. M. (1982). A quantitative study of thalamocortical and other synapses involving apical dendrite of corticothalamic projection cells in mouse smI cortex. *J. Neurocytol.* 11:137–57.

White E. L. and Rock M. P. (1981). Comparison of thalamocortical and other synaptic inputs to dendrites of two non-spiny neurons in a single barrel of mouse smI cortex. *J. Comp. Neurol.* 229:311–20.

3
Processing of spikes by neural networks

3.1 Introduction

The preceding chapter dealt with methods for calculating the probability of finding synaptic contacts between neurons. In this chapter we assume that such contacts exist and create a certain network. This chapter develops methods for computing the input–output relations of a network from its structure.

Chapters 2 and 3 provide the material essential for analyzing small cortical networks. It is asumed that one can experimentally observe the input–output relation of a cortical network. Then, with the insight gained from studying Chapter 3, one can surmise the structure of cortical networks that might generate the observed relation. Finally, using the methods described in Chapter 2, one can determine which of the possible networks is most likely to exist in the brain. Chapters 6 and 7 will illustrate an important example of the use of such considerations.

This chapter derives quantitative descriptions of the way in which a spike train arriving at a neural network is modified and converted into an output spike train. The reader should be aware that this assignment is very ambitious and can be only partially achieved. Specifically, we could arrive at formulas that would hold true for spike trains in which the spikes did not follow each other too closely (we would, of course, have to define what "too closely" means). In this chapter we deal extensively with spike trains of one axon, although a realistic view of cortical transformations must take into account the fact that the usual input into a cortical net is composed of a whole ensemble of spike trains arriving through many parallel inputs, and, similarly, the output of a cortical network should be described in terms of parallel spike trains. These issues are dealt with partially in this chapter and will be taken up again (although only for certain limited types of parallel lines) in Chapter 7.

Throughout this chapter, we assume that we do not have any information regarding intracellular events. That is, we can measure only extracellularly recorded spike trains. With that limitation in mind, we start our discussion with a question: How does a single spike from the presynaptic axon affect the postsynaptic spike train? In dealing with this issue, we must define the concept of the "gain" of a synapse. Following the definition of synaptic gain, we devote a short section to ways in which this concept can be used. Then we elaborate on the topic of transmission through a single synapse in order to derive a quantitative description of the transmission of a whole spike train from a presynaptic axon to a postsynaptic neuron. We conclude this chapter by illustrating how the formulas derived for transmission through a single synapse can be used to work out the transfer functions for neural networks, including neural circuits with recurrent connections (feedbacks).

A concept that is central to a description of the firing times in a spike train is the "instantaneous firing rate." The remainder of this section deals with the main features of this descriptor of spike trains.

Let us assume that an axon or several parallel axons arrive at a cortical network. The input arriving through these axons is processed in some fashion by the network and then output through another axon or several parallel axons. If we wish to describe such a system quantitatively, we must first find a quantitative manner to describe the spike trains that come into the network or go out of the network.

The first possibility that may come to mind is to describe these spike trains by specifying the times of occurrence of the spikes in the train $(t_1, t_2, t_3, \ldots)$; but even if we specify such times for the input train (e.g., 0.005, 0.27, 0.54, $\ldots$), it will be impossible to determine the exact times for the output train because there are many other variables in the brain (such as the firing times of other, unknown, neurons) that affect the output. Therefore, the output train will depend on the input train only in a probabilistic sense. In most instances, the input train cannot be controlled to the extent of being able to specify the exact list of firing times, because the input axon or axons cannot be directly stimulated. If, for instance, we generate the input to the network by giving some external sensory stimulus, we are able to define the resulting input spike train only in probabilistic terms.

The simplest way to describe the spike train in terms of probability is to refer to its firing rate λ (say five firings per second). This is converted into probability in the following way: If we choose a very small time interval Δt, then the probability of finding a spike between t and $t + \Delta t$ is $\lambda \Delta t$. (This becomes absolutely true only when Δt is so small that the probability of getting two spikes within Δt is zero.) This relation is stated as

$$\text{pr\{a spike in } (t, t + \Delta t)\} = \lambda \cdot \Delta t \tag{3.1.1}$$

this means that if we attempt many (N) times to determine if there is a spike in $(t, t + \Delta t)$, then we expect to succeed in finding a spike in $N \cdot \text{pr}\{\}$ times, as in this formula:

$$E[\text{number of successes}] = N \cdot \text{pr\{a spike in } (t, t + \Delta t)\} \tag{3.1.2}$$

where the term $E[\ ]$ means the expected value of the variable in square brackets. We could get as close as desired to the ratio $E[\text{number of successes}]/N$ if we made a large enough number of measurements. We note that equation (3.1.2) is accurate only for very small Δt. But what can we say if we wish to consider a larger Δt, as is often done when constructing histograms from recorded data? In such a case, we cannot speak of the number of successes, but rather of the number of spikes. The most intuitive relation, then, is that the expected number of spikes during T seconds is $\lambda \cdot T$. This is, in fact, the definition of rate. The relation

$$\text{number of spikes} = \lambda T$$

also holds true when T is not one continuous time period, but is made up of many (N) little sections of length Δt. We suggest, therefore, that an equation similar to equation (3.1.2) for the number of spikes (instead of number of successes) is

$$E[\text{number of spikes}] = N \cdot \lambda \cdot \Delta t \tag{3.1.3}$$

For the smallest nontrivial case, in which we have only one trial ($N = 1$), we arrive at an equation similar to equation (3.1.1):

$$E[\text{number of spikes}] = \lambda \cdot \Delta t \tag{3.1.4}$$

The last two relations (where we deal with the number of spikes rather than the number of successes) are true for any size of Δt. The difference is due to the fact that numbers of spikes are additive, whereas probabilities are not. Therefore, if we take one long Δt and break it into a number of smaller contiguous sections, the expected number of spikes in the long Δt will be the sum of the expected number of spikes from all the smaller sections, whereas the probability of a success in the long Δt (i.e., the probability of getting one or more spikes in Δt) is *not* the sum of all the probabilities of getting a spike in the smaller sections. We should note that when Δt is small enough (i.e., so small that the probability of getting two spikes within Δt is negligible), then "number of spikes" is equal to "number of successes."

Throughout the preceding discussion, the firing rate (λ) was treated as if it had a constant value throughout time, but there is no reason for that limitation. The equations speak only to what happened in the interval from t to $t + \Delta t$. Had λ assumed different values before t or after $t + \Delta t$, it would not have affected the equations at all. Therefore, when Δt is small enough, we can talk about the firing rate as a function that varies with time: $\lambda(t)$. An infinitesimally small Δt is not a practical entity to use for real data, but if $\lambda(t)$ changes slowly within Δt, we can still apply each of the foregoing formulas for a firing rate that varies through time. We can now restate equations (3.1.1) through (3.1.4) in terms of variable firing rates. For brevity, the probability of finding a spike between t and $t + \Delta t$ is denoted by P; the number of successes (i.e., finding at least one spike in the stated interval) in N trials is denoted by nss; the number of spikes (recorded during the stated interval) in N trials is denoted by nsp:

$$P = \lambda(t)\Delta t \ \text{(for small } \Delta t) \tag{3.1.5}$$
$$E[\text{nss}] = NP \tag{3.1.6}$$
$$E[\text{nsp}] = N\lambda(t)\Delta t \tag{3.1.7}$$

The function $\lambda(t)$, which describes the variations in the firing rate of an axon as a function of time, is called the instantaneous firing rate; see Cox and Isham [1980] for a rigorous treatment of this term.

In order to get a better intuitive feeling for the notion of instantaneous firing rate, we can compare it to the speed of a car. If we are driving at a constant speed, the distance we travel can be determined by multiplying the speed by the time of travel. If the driving speed varies, we should integrate the speed over time to obtain the distance. Similarly, if the firing rate is constant, we can determine the expected number of spikes in a given time interval by multiplying the rate by the measuring time, but if the firing rate varies, we must integrate the firing rate over time to determine the expected number of spikes.

What use do these formulas have in real life? There are two situations in which they can be used: (1) for making experimental checks on the prediction of models and (2) for estimating the parameters of recorded data.

Prediction validation may be conducted in the following manner: A model should tell us the expected rate function $\lambda(t)$. We can test this predicted rate against the measured data by counting the number of spikes observed in N trials during the time $(t, t + \Delta t)$ and comparing the observed count to the count we expected:

$$E[\text{nsp}] = N \int_{t}^{t+\Delta t} \lambda(s) \, ds \tag{3.1.8}$$

Although the preceding argument was formulated for checking the prediction at one particular time (t), it is true for whatever value we assign to t. Therefore, by measuring N responses, we can affirm the prediction for all t values; i.e., we can test the predicted time course of λ experimentally.

Parameter estimation proceeds in a complementary fashion. For every time (t), we count the number of spikes (nsp) occurring in N trials during the time sections $(t, t + \Delta t)$, and then we estimate the rate function in that t by

$$E\lambda(t) = \text{nsp}/(N\Delta t) \tag{3.1.9}$$

The estimate is accurate in the following sense. If we could repeat the measurement as many times as we wished, the average computed firing rate $E\lambda(t)$ would be as close as we wished to the true rate at some point between t and $t + \Delta t$. This last statement can be written

$$E[\text{nsp}/(N\Delta t)] = \lambda(t + \theta\Delta t), \qquad 0 \le \theta \le 1 \tag{3.1.10}$$

The reader probably has noticed that what is known in the physiological literature as PSTH (peristimulus time histogram) is an estimate of the instantaneous firing rate, except perhaps for a scaling factor of the ordinate [Gerstein and Kiang, 1960].

To summarize, the instantaneous firing rate $\lambda(t)$ is a descriptor for the firing pattern of an axon, and it lends itself to analytic calculations, to verification in experiments, and to estimation by measurements.

In the following section, we make extensive use of the concept of instantaneous firing rate. First we examine the instantaneous firing rate of the postsynaptic cell as a function of time after the firing of a presynaptic axon.

3.2 Transmission of a single spike through a synapse

Theoretically, we could measure the input–output relationships of a synaptic connection by performing the hypothetical experiment described in Figure 3.2.1. The experiment is conducted as follows: At random times we stimulate axon a_1 and measure the response coming

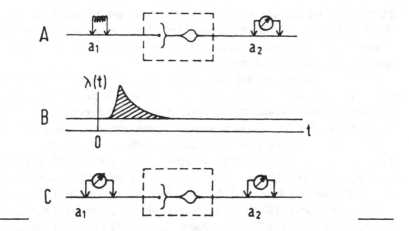

Figure 3.2.1. Measuring synaptic transmission. A: Experimental setup for measuring. B: The synaptic transmission curve. C: Estimating the synaptic transmission curve from the spontaneous activity.

through axon a_2. After many repetitions of this experiment, we should be able to draw a graph showing the firing rate of axon a_2 as a function of the time that passed after we stimulated a_1.

Ideally, such a graph might look like Figure 3.2.1B. Most cortical cells fire at some nonzero rate even without any known external stimulus, because every cortical neuron receives inputs from thousands of axons, most of which are active at seemingly random times. This so-called spontaneous rate shows up as a constant level to which the synaptic transmission curve is added. The curve departs from this constant level at some time after the stimulus, for several reasons: because the action potential elicited by our stimulus (in a_1) has to travel along the axon to the synaptic site, because there is a synaptic delay during which the effect of the presynaptic spike is transmitted to the postsynaptic site, and because when a postsynaptic spike is generated, it takes some time until it arrives at the recording electrodes. Usually, if the stimulating and recording electrodes are not far away from the synapse, the delay will be in the range of 1–2 ms. (The delay to an intracellularly recorded synaptic potential is 0.5–1 ms, and the delay until the synaptic depolarization produces an action potential is another 0.5–1 ms.)

The shape of the transmission curve obtained in this manner depends on, but is not identical with, the shape of the intracellularly recorded synaptic potential (this issue is taken up extensively in Chapter 4). For

the moment, we are interested only in the input–output relations of the
"black box" that includes the synapse between a_1 and the dendrites (or
soma or axon) of the cell from which a_2 emanates, not in the intracellular
mechanisms underlying these relations.

A useful descriptor of the synaptic strength can be obtained by integrating the synaptic transmission curve. The area between the level of
the spontaneous rate and the synaptic transmission curve (the hatched
area in Figure 3.2.1B) tells us how many action potentials will be added,
on the average, to the output stream per each action potential in the
input. This area is the gain of the synapse measured while activity in the
input was not synchronized with any other inputs to the postsynaptic
cell. This gain will be called the *asynchronous gain* (ASG). (Levick,
Cleland, and Dubin [1972] called this area the "effectiveness of a connection between two cells.") The response curve will be called the asynchronous synaptic transmission curve.

3.2.1 Practical considerations

Usually it is not possible to conduct the experiment described in
Figure 3.2.1A because of the great difficulty in stimulating a single axon
in the cortex. Therefore, one usually conducts the experiment described
in Figure 3.2.1C, in which both the presynaptic and postsynaptic spike
trains are recorded. The firing rate of axon a_2, as a function of time
elapsed since the firing of a_1, is computed from the recorded data. This
computed function is known in the physiological literature as the cross-
correlation function [Perkel, Gerstein, and Moore, 1967] or the cross-
renewal density function [Abeles, 1982a]. The correct statistical name
probably should be conditional intensity function or cross-conditional
intensity function [Cox and Isham, 1980, p. 13], but we shall use the
term favored by most neurobiologists: cross-correlation.

The cross-correlation between spontaneously active nerve cells will be
affected by the structure of the presynaptic spike train. Mathematically,
the only case for which the cross-correlation will be identical with the
synaptic transmission curve is that in which the firing times of the input
axon (a_1) are not correlated to the firing times of all the other inputs that
affect a_2 and when the firing times of the input obey the statistics of a
uniform Poisson process (i.e., the probability of finding a spike is independent of the time and of the past history of the axon).

These two requirements are extremely restrictive, but the applicability of the cross-correlation can be increased under the following circum-

stances: If the input spike train is non-Poissonian, but behaves as a renewal process (i.e., the possibility of finding a spike depends only on the time elapsed since the last spike [Cox, 1962]), then the true synaptic transmission curve can be obtained by deconvolving the apparent response curve from the autocorrelation curve of the input train [Perkel et al., 1967].

The requirement that the firing times of the input axon be uncorrelated with any other input cannot be met, because approximately half of the pairs of cortical neurons show a weak correlation that is bound, to some extent, to distort the measured synaptic transmission curve. This distortion usually is ignored because the details of the correlations between cortical neurons are not known. Many of the correlations between pairs of cortical neurons show positive peaks; negative peaks are very rare. Therefore, the transmission curve obtained from correlating spontaneous activity is higher than one would expect to obtain for an uncorrelated input.

The requirements of a constant firing rate of the presynaptic neuron and a lack of correlation between its firing times and the other presynaptic sources are strongly violated when an external stimulus is applied. When a stimulus is applied, many neurons vary their firing times in a coordinated fashion. Yet we often wish to obtain synaptic transmission curves (by computing the cross-correlation function) when a stimulus is present. It is possible to evaluate the amount of correlation introduced by the stimulus itself by computing the so-called shift-predictor correlogram [Perkel et al., 1967; Dickson and Gerstein, 1974]. This is achieved by taking the firing times of the presynaptic neuron in relation to the onset of one stimulus presentation and the firing times of the postsynaptic neuron in relation to the onset of another stimulus presentation and then computing the cross-correlation. Any correlation that is directly introduced by the stimuli will show up in the shift-predictor cross-correlogram. A method for predicting how the cross-correlation function introduced by a stimulus will vary with the time after presentation has recently been suggested [Palm, Aertsen, and Gerstein, 1988; Aertsen et al., 1989]. This method is known by the name "joint poststimulus time histogram" (JPSTH).

In most cases, when measuring the correlation between two cortical cells, we have no a priori notion regarding which cells are connected to what. The argument given earlier usually is inverted so as to claim that when we find a cross-correlation that looks like Figure 3.2.1B, there is a synaptic interaction between the two cells, and the cross-correlation ob-

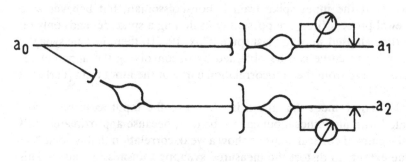

Figure 3.2.2. An alternative explanation for the situation in which a_2 fires at a higher rate after a_1 has fired.

tained (after normalization of the ordinate to give spikes per second) is the synaptic transmission curve. However, it must be remembered that we cannot infer causal relationships from correlations. There are alternative mechanisms that can produce graphs such as that in Figure 3.2.1B without a direct synapse between the two cells. An example of such an alternative mechanism is shown in Figure 3.2.2.

Later we treat in greater detail the question of the difference in cross-correlations between cells a_1 and a_2 when we compare the connection in Figure 3.2.1A to the network of cells in Figure 3.2.2. Here it suffices to note that unless the synapse between a_0 and the interneuron in Figure 3.2.2 is very strong, the cross-correlation will be extremely weak.

In short, it is practical to measure the response curve by measuring the cross-correlation between two cells. Whenever there is a synapse between the two cells, we expect to find a curve like that in Figure 3.2.1B, though the inverse is not true. Therefore, when one conducts an experiment in which the activities of pairs of neurons are correlated, the number of pairs for which such a curve is found will be greater than the number of pairs that are actually connected by a synapse. One must also remember that for a finite recording time, there will always be a minimal synaptic strength below which synapses cannot be detected by cross-correlation.

Exercise 3.2.1. We record extracellular activities from two neurons firing at a rate of five spikes per second. It is assumed that there exists an excitatory synapse between the neurons whose ASG is 0.01 and whose effect lasts 0.01 s. For how long do we need to measure the activities of the two neurons if we wish to assert (with probability

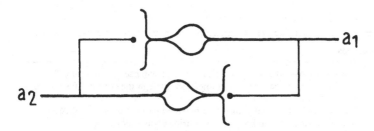

Figure 3.2.3. Mutual synapses.

of error less than 0.01) that indeed there is a synapse between the cells?

Exercise 3.2.2. The activities of two neurons were recorded for 20 min. Each of them fired at five spikes per second. What is the weakest ASG that can be detected by counting the excess of spikes that one cell fired during 0.01 s after the other cell fired?

Exercise 3.2.3. Two cells establish synaptic contact with each other, as shown in Figure 3.2.3. How will the synaptic transmission curve look when measured by the method described in Figure 3.2.1A (direct stimulation)? How will it look when measured by the method of Figure 3.2.1C (cross-correlation of spontaneous activity)? Draw qualitative graphs, and explain the conclusions.

3.3 Using the ASG

In the preceding section we defined the ASG as the area between the synaptic transmission curve and the level that represents the spontaneous firing rate of the postsynaptic cell (the hatched area in Figure 3.2.1B). The ASG expresses the average number of extra spikes that are added to the postsynaptic train for each presynaptic spike. This section presents some examples of how this gain can be used to assess the functioning of various simple networks.

3.3.1 Experimental evaluations of the ASG

The ASGs in the cortex have values much smaller than unity. This can be seen from various data published in the literature and sum-

Table 3.3.1. *Values of ASG in the cortex*

Methods	ASG
1. Theoretical average ASG for an anatomical synapse	0.003
2. Theoretical average ASG for connection through the corpus callosum	0.025–0.013
3. Average ASG for local connection between pairs of adjacent neurons in the auditory cortex during spontaneous activity	0.06
4. ASG for two auditory neurons during spontaneous activity	0.015
5. ASG for the same pair as in 4, but during acoustic stimulation	0.09
6. ASG for two auditory neurons during spontaneous activity	0.14
7. ASG for two visual neurons during photic stimulation	0.25
8. ASG for two visual neurons during photic stimulation	0.146
9. ASG for two visual neurons during spontaneous activity	0.086
10. ASG for the same pair as in 9, but during photic stimulation	0.24

Sources: 1, from Abeles [1982b]; 2, from Braitenberg [1978a]; 3, from Abeles [1982b]; 4, 5, and 6, computed from Figures 11E, 11H, and 4C of Dickson and Gerstein [1974]; 7 and 8, computed from Figures 1A and 1B of Michalsky et al. [1983]; 9 and 10, computed from Figures 5D and 5F of Toyama et al. [1981].

marized in Table 3.3.1. In the study reported as entry 1 of Table 3.3.1, an attempt was made to assess the strength of what seemed to be one anatomical synapse. For all the other entries, the strength of connection between two cells was estimated. However, two cortical cells may be connected by more than one anatomical synapse. The estimates in entries 4 through 10 were based on individual figures for synaptic connection published as cross-correlations in the literature. Such examples tend to show the strongest effects seen. Therefore, these values should be taken as the upper limits of the values that may be found in the cortex.

In the following chapters, we will need an estimate for the range of synaptic strengths that one might expect to find. We shall assume that they fall in the range of 0.003 to 0.2, where only a small fraction of synapses will approach the 0.2 limit. It should be emphasized that these figures relate to the strength of the local connection between two cortical neurons. Other types of connections (such as may be formed between thalamic input fiber and a cortical cell) may be much stronger.

The preceding discussion has dealt with excitatory connections, but inhibitory connections can also be seen in cross-correlations. They would be represented as a trough in the synaptic transmission curve,

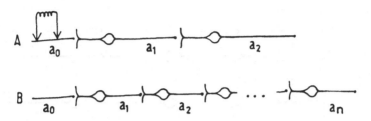

Figure 3.3.1. Transmission through a chain of synapses. A: Two synapses in tandem. B: *n* synapses in tandem.

and they would have negative ASG values. However, it is difficult to assess their strength because the literature contains very little information on them, for a variety of reasons. Most cortical cells have very low firing rates, and therefore a great amount of information is needed before an inhibitory effect can be shown [Aertsen and Gerstein, 1985]. In addition, inhibitory cells may have low spontaneous firing rates, and it may be difficult to pick up the activity of inhibitory cells using metal microelectrodes. Be it as it may, the lack of data in the literature prevents us from estimating the strengths of inhibitory synapses in the cortex.

Therefore, our numerical examples will treat only excitatory connections. However, all the arguments and equations should be applicable to inhibitory connections (with negative ASG) as well.

3.3.2 Transmission through a chain of synapses

Let us assume that we are facing a chain of excitatory connections, as shown in Figure 3.3.1A, in which the gain of the first synapse is ASG_1, and the gain of the second synapse is ASG_2. For each input spike (at axon a_0) we then obtain an average of ASG_1 spikes in axon a_1, and for each additional spike in that (a_1) axon, we then get an average of ASG_2 additional spikes in axon a_2. Therefore, for each spike in a_0, $ASG_1 \cdot ASG_2$ spikes are added to the spike train of a_2.

In general, if we face a series of *n* synapses, such as in Figure 3.3.1B, in which the gain at the *i*th synapse is ASG_i, the overall gain between the input axon a_0 and the output axon a_n will be

$$ASG = \prod_{i=1}^{n} ASG_i \qquad (3.3.1)$$

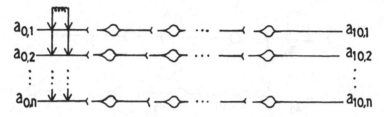

Figure 3.3.2. Transmission through parallel chains.

Example 3.3.1. Assume that a central process that takes 100 ms is conducted through 100 nerve cells that are organized in series. Then, even if the synapses are all of the strongest type (ASG = 0.2), the transmission gain through the chain is

$$0.2^{100} \approx 1.3 \cdot 10^{-70}$$

It is clear that such a chain cannot transmit single spikes. Even if the chain were only ten synapses long, its gain would be $0.2^{10} \approx 10^{-7}$, which can, for all practical purposes, be considered as zero again.

Exercise 3.3.1. One usually assumes that transmission through the cortex is achieved through many parallel channels. Let us assume that we have a network of many (n) parallel chains, each of which is ten synapses long (this system is described in Figure 3.3.2). Assume that we stimulate all the input axons simultaneously and that all the synapses have the strength (ASG) of 0.2. How many parallel channels (n) do we need if we wish to get, on the average, at least one extra spike on one of the output axons for each stimulus?

3.3.3 The effect of a single cell on a population of cells

Each pyramidal cell in the cerebral cortex establishes synaptic contacts with many other cells. Some of these are in the neighborhood of the presynaptic cell body, and others are found in distant regions (reached through that part of the axon that passes through the white matter). This situation is shown schematically in Figure 3.3.3.

Let us denote the average synaptic gain of axon a_0 by ASG. Because a_0 contacts n cells, we expect that for each action potential in a_0 there should be an average of nASG spikes in the entire postsynaptic population.

In the cortex we estimated that the average strength could be as weak

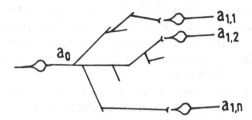

Figure 3.3.3. One pyramidal cell excites n other cells.

as 0.003. Each pyramidal cell will give synaptic contacts locally to about 5,000 of its neighbors. Therefore, each action potential in a pyramidal cell will evoke, on the average, fifteen $(5,000 \cdot 0.003)$ spikes in its neighboring cells. This effect seems quite strong, but we must remember that this is the marginal gain of the synapses. That is, on the background of the ongoing cortical activity, each additional spike in a pyramidal cell will evoke fifteen additional spikes in its neighbors.

The remainder of this section attempts a rough calculation of the consequences of this (1 : 15) sensitivity. Throughout this discussion, we assume that time propagates in discrete steps. Each step is one synaptic delay long (which we assume is 1 ms). We also assume that all the synaptic effects last for one time step only. These assumptions are very crude, but they simplify the calculation of estimates of the development of activity through time. (Much of the theoretical research done on neural networks uses such simplifying assumptions.) While it is clear that such assumptions yield only approximately correct results, they are still useful because they provide a feeling for the types of behaviors that we can expect from neural networks.

Let us assume that we have a cortexlike network composed only of excitatory cells (this situation can be achieved experimentally by blocking the inhibitory transmitters with drugs). Such a cortical network includes approximately 10^{10} cells, each of which is connected to about 5,000 neighboring cells (and another 5,000 distant cells). On the average, each cell fires five times per second. This means that for every time step (1 ms), we have an average of approximately 50 million spikes $(5 \cdot 10^{-3} \cdot 10^{10})$ in the entire network. The actual number fluctuates around this average, with a standard deviation of about 7,000 $(50,000,000^{1/2})$. Let us investigate what would happen if, at a given millisecond, the number of firing cells were one standard deviation above that average (i.e., instead of 50 million spikes we had 50,007,000 spikes). The extra

7,000 spikes would evoke a surplus of 100,000 spikes (15 · 7,000) after one synaptic delay. These would, in turn, evoke 1.5 million spikes after one additional synaptic delay, which would evoke 22 million superfluous spikes after yet another synaptic delay. Within about two additional delays, the entire cortex would be firing.

In fact, the explosion of activity would be much faster, because of three factors: We did not take into account the effects that are conducted through the white matter (which require a little longer to arrive at the target cells). We also failed to take into account the effect of the background excitation on the synaptic gain (Chapter 4 shows that the ASG increases with cell depolarization). Furthermore, we did not consider the effect of synchronized activity, which starts to dominate when the firing rate increases (Chapter 7 shows that the gain of a synchronous volley of presynaptic spikes is larger than the sum of the asynchronous gains of the constituent spikes).

The inverse process can be expected when the firing rate is one standard deviation below the average: Within one synaptic delay, 100,000 spikes would be missing, then 1.5 million would be missing, and so forth. Within a short time the entire cortex would become quiet.

Although these calculations are extremely rough, they do show an important basic property of excitatory networks: In a purely excitatory network, partial activity cannot be maintained in a steady state. Even if the activity in the network remained at a steady level momentarily, it would decay to zero within a short time and stay there, or it would reach saturation (i.e., most of the neurons either firing or refractory) and stay there. We shall expand our discussion of this instability in Chapter 5. The simplified argument of the preceding paragraphs also shows that the decay (or explosion) of activity is so fast that if one wants to stabilize the network's activity with a negative feedback, one must take great care to build a feedback with the smallest possible delays.

This gives us some insight into the cortical arrangement in which the cortical inhibitory interneurons overlap with the excitatory ones and exert much of their inhibitory effects in the same local region from which they derive their excitation. Many of these inhibitory neurons have relatively thick and partially myelinated local axons. All these features guarantee a stabilizing feedback with the minimal possible delay. The speed of explosion (or decay) of activity in an excitatory network is so great that there is not much sense in having an inhibitory feedback arrive at the cortex from deeper nuclei. It would take 10 ms or more before the feedback loop could sense a change in the cortical level of activity and attempt to correct it by increasing (or decreasing) the inhibitory level of the cortex. By that time, the cortical activity would already have reached an extreme level of full activation (or silence). Indeed, we know of no inhibitory inputs that reach the cortex through the white matter.

Exercise 3.3.2. We are presented with a network having two types of neurons: 80 percent of the cells are excitatory, and 20 percent are inhibitory. Every excitatory cell makes random contacts with N cells (excitatory or inhibitory). All the contacts have the same gain E. Every inhibitory cell makes contacts with N cells, all of which are excitatory; that is, an inhibitory cell may not inhibit another inhibitory cell. The strength of the inhibitory connections is I. In its normal state, the network has a steady state of activity in which every cell is active at some low level.

Let us assume that time can be divided into 1-ms steps, that the synaptic delay is 1 ms, and that the postsynaptic effect lasts for only 1 ms. We wish to ensure the following property: If at a certain millisecond the number of active excitatory cells increases slightly, then the extra inhibition that will be generated by that increase 2 ms later will exactly balance off the extra excitation at the time. Under these circumstances, determine the ratio I/E.

Problem 3.3.1. If one takes into account excitatory and inhibitory effects that last more than 1 ms, one discovers that the level of activity shows strong oscillations between high and low levels. Find ways to build a network of excitatory and inhibitory cells that will show a stable low level of spontaneous activity.

3.3.4 The effect of a population of cells on one cell

In the preceding section we discussed the effect of one cell on a population of postsynaptic cells. We now wish to discuss the inverse situation in which an entire population of presynaptic cells converge on one postsynaptic cell. This situation is described schematically in Figure 3.3.4. Let us denote the gain of the synapse between the ith axon and the postsynaptic cell by ASG_i. Then, if the cells in the presynaptic population are firing at a slow asynchronous rate, we can assume that the probability that two of the cells will fire simultaneously is negligible. We can then define the gain of transmission between the population and the postsynaptic cell as the average number of spikes in the postsynaptic cell

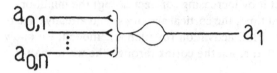

Figure 3.3.4. n axons converge on one postsynaptic cell.

evoked by a spike in one of the cells in the presynaptic population. Under such conditions we have

$$ASG = (1/n) \sum_{i=1}^{n} ASG_i \qquad (3.3.2)$$

We note that when we have a chain of synapses in tandem (Section 3.3.2), the overall gain is dependent on the product of the individual gains, whereas when we have parallel synapses, the overall gain depends on the sum of all the gains.

The requirement that all the presynaptic cells fire in complete asynchrony is quite restricting. If we wish to deal, in addition, with cases in which two or more cells are allowed to fire together, we must define the transmission gains for such occurrences. This is important, because a few concurrent presynaptic spikes may have a postsynaptic effect much stronger than the combined effects of the same spikes arriving asynchronously (Chapter 7 deals extensively with this phenomenon).

In the following discussion, we assume that all the n parallel synapses have the same properties. Let us denote by $\mathrm{pr}\{n\}$ the probability that n cells will fire together at a given millisecond, and by $ASG(n)$ the gain of a single presynaptic spike when it arrives together with other $n - 1$ presynaptic spikes. Then we have the following situation: When only one spike arrives, it will generate an average of $ASG(1)$ spikes in the postsynaptic cell. When two spikes arrive simultaneously, they will generate an average of $2ASG(2)$ spikes in the postsynaptic cell [we multiply the ASG by 2 because each of the two presynaptic spikes has a gain of $ASG(2)$]. When k presynaptic spikes arrive together, they will generate an average of $kASG(k)$ spikes in the postsynaptic cell.

If we know all these gains and the probability of their occurrence, then we can estimate the generalized transmission gain between the population and the postsynaptic cell as

$$ASG = \sum_{i=1}^{n} \mathrm{pr}\{i\} i ASG(i) \qquad (3.3.3)$$

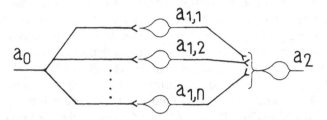

Figure 3.3.5. Diverging and converging connections between cells.

Exercise 3.3.3. A population of 200 cells is connected to one postsynaptic cell, as shown in Figure 3.3.4. Each of the presynaptic cells fires at an average rate of five spikes per second. Firing times are independent, and so the probability of k spikes arriving at the same millisecond is given by the Poisson distribution:

$$\text{pr}\{k; x\} = e^{-x}x^k/k!$$

where x is the expected number of arriving spikes (within 1 ms). Let us assume that the gain of a single synapse is 0.01 and that when k synapses arrive together their combined strength is given by

$$k\text{ASG}(k) = 0.01 \cdot 2^{(k-1)}$$

What will be the average gain of transmission between the population and the postsynaptic cell?

Exercise 3.3.4. One axon (a_0) branches off and makes excitatory contacts of strength ASG with n interneurons. These interneurons then converge and establish excitatory synapses of the same strength to a postsynaptic cell (a_2). This situation is shown in Figure 3.3.5. For simplicity, let us assume that the synchronous firing of the presynaptic cells does not affect the synaptic gain. What will be the transmission gain between a_0 and a_2?

Exercise 3.3.5. For the same connections as in Figure 3.3.5, let us assume that there are 100 parallel interneurons between a_0 and a_2 and that the gains are given by

ASG(1) = 0.01
ASG(2) = 0.02
ASG(3) = 0.03, etc.

What is the transmission gain between a_0 and a_2?

Problem 3.3.2. In Section 2.3 we calculated the probability of
the existence of connections between adjacent cortical cells. According
to Braitenberg [1978b], the local axonal branches of a pyramidal cell are
distributed below its cell body, and the basal dendrites branch out approximately at the level of the cell body. Therefore, if we select two
pyramidal cells at the appropriate distance, we have the situation shown
in Figure 3.3.5, where the top pyramidal cell is a_0, below it is a population $(a_{1,i})$ of pyramidal cells, and below that is the other (a_2) pyramidal
cell we select.

In order to decrease the number of possibilities, let us assume that the
axon of the pyramidal cell is distributed in a sphere whose center is 100
μm below its cell body, that the basal dendrites of the pyramidal body
are centered around the body, and that their density falls exponentially,
as described in Section 2.3.2. For simplicity, let us also assume that the
effects of synchronously firing cells add up linearly [i.e., $ASG(n)$ =
$ASG(1)$]. If a spike is generated by the top pyramidal cell, what will be
the strength of its effect on the bottom cell when the two pyramidal cells
are positioned one below the other at the optimal distance?

3.4 Transmission of a spike train

Until this point, we have investigated transmission through synapses by
counting the number of spikes that are added to (or deleted from) the
output spike train. In this section we also examine the temporal development of the effects of transmitting through a synapse. Figure 3.4.1 shows
a schematic case in which such time–structure considerations are useful.
This figure describes the situation in which the firing rate of the
presynaptic axon a_0 varies over time because of the effect of an external
stimulus that is applied to the nervous system. Assuming that the response curve of the presynaptic fiber $[\lambda_0(t)]$ and the synaptic transmission curve $[h(t)]$ are known, we can determine how the postsynaptic cell
a_1 will respond to the stimulus.

Let $\lambda_1(t)$ denote the response of the output axon a_1 to the stimulus,
and let us assume that for all the curves $(\lambda_0, \lambda_1, h)$ the spontaneous firing
level was subtracted. We claim that if the firing rate λ_0 does not reach
high levels [at which $h(t)$ changes], the output response is obtained by a
convolution between the response curve of the input axon and the
synaptic transmission curve. The equation then will read

$$\lambda_1(t) = \lambda_0(t) * h(t) \tag{3.4.1}$$

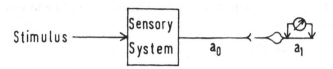

Figure 3.4.1. Transmission of a response through a synapse.

where the asterisk means the convolution integral

$$f(t) * g(t) = \int_{-\infty}^{\infty} f(t - \tau)g(\tau) \, d\tau$$

Proof. Let us first examine what occurs at a particular time instant (τ) in the input axon (a_0) and what the effect may be on the firing of the output axon at another time instant (t). Let us assume that we repeat the observations at these two instants, while presenting the external stimulus repeatedly for N times. As was described in Section 3.1, the expected number of spikes (n) in the presynaptic axon at the time bin (τ, $\tau + d\tau$) is given by

$$dn = N\lambda_0(\tau) \, d\tau \tag{3.4.2}$$

Each of these added spikes in the presynaptic axon may add a spike to the postsynaptic train at time (t, $t + dt$). Let dm denote the expected number of these added spikes. Then we expect to get

$$dm = dn \, h(t - \tau) \, dt \tag{3.4.3}$$

If we substitute dn in equation (3.4.3) by equation (3.4.2), we arrive at

$$dm = N\lambda_0(\tau)h(t - \tau) \, d\tau \, dt \tag{3.4.4}$$

However, if we wish to determine the total number of additional spikes in the output axon at (t, $t + dt$), we must consider all possible τ values at which the input axon can contribute to the output stream. That is, we must consider what might happen at the input axon for all τ values between zero (the time of stimulation) and t (the time of measurement of the output). The effects at all these τ values will be additive as long as the synaptic response curve $h(t)$ is a valid descriptor of the transmission by the synapse. (The latter requirement was one of the assumptions we made. This issue will be discussed again after the proof of our claim is completed.) By summing up all these effects, we arrive at

$$m = N \, dt \int_0^t \lambda_0(\tau) \cdot h(t - \tau) \, d\tau \tag{3.4.5}$$

The definition of the response curve of a neuron to a stimulus was given in Section 3.1 as

$$m = N \, dt \, \lambda(t)$$

By comparing equation (3.4.5) with this definition of the response curve, we arrive at

$$\lambda_1(t) = \int_0^t \lambda_0(\tau) \cdot h(t - \tau) \, d\tau \tag{3.4.6}$$

This equation differs from our original claim only in the limits of integration, which had been $(-\infty, \infty)$ in the claim, and are $(0, t)$ here. However, both $h(t)$ and $\lambda_0(t)$ have zero values for negative times (i.e., we make the reasonable assumption that the systems are *causal*). By extending the integration limits to $-\infty$, we do not change the value of the integral [because $\lambda_0(\tau)$ is zero for negative τ]. Extending the upper limit of integration to $+\infty$ will not change the value either [because $h(t - \tau)$ becomes zero when τ is larger than t]. This concludes our proof.

Exercise 3.4.1. Prove the same claim by using probabilities of firing rather than instantaneous firing rates.

Let us now return to the requirement that the firing rate in the presynaptic axon $[\lambda_0(t)]$ not be too high. One must remember that if two presynaptic spikes are too close together, the response to the second spike may differ from the response to the first (departing from the assumption of linearity). Thus, for instance, if $h(t)$ is significantly different from zero for 5 ms, we should not try to apply equation (3.4.1) for responses in which there is a high probability of two spikes occurring within 5 ms at the presynaptic axon (a_0). If, for instance, we want to ensure that pr{2 spikes within 5 ms} < 0.1, we should find x such that

$$\text{pr}\{2; x\} < 0.1$$

For a Poissonian spike train, that means

$$e^{-x} \cdot x^2/2 < 0.1$$

which means that $x < 0.6$. That is, at its peak, $\lambda_0(t)$ should not exceed $0.6/5 \cdot 10^{-3} = 120$ spikes per second.

The conclusion thus arrived at is that as long as we do not reach high firing rates (above 100 spikes per second), the synapse behaves like a linear system, and the synaptic transmission curve can be understood as the impulse response of that system.

This conclusion may at first seem improbable, considering that the cellular mechanisms for generating an action potential are extremely nonlinear. In Chapter 4 we investigate synaptic transmission from an intracellular point of view. There we show that the random fluctuations of the membrane potential (caused by continuous impingement of thousands of excitatory and inhibitory effects) are responsible for the "linearization" of the input–output relations of the synapse. But before turning to that issue, we wish to exploit the apparent linear behavior of the synapse.

To summarize, when firing rates are below 100 spikes per second, the synapse acts like a linear filter. The gain and the characteristics of the synapse can be estimated from the cross-correlation between the firing times of the presynaptic and postsynaptic neurons. Using these characteristics, one can compute the input–output relations of more complex neural circuits by using linear systems analysis, a powerful technique used extensively in engineering. The reader interested in pursuing the subject can find many textbooks in this field, such as that by Siebert [1986]. Here we attempt to illustrate its potential through the examples in the next section.

3.5 Examples

In this section we illustrate the use of the conclusion from Section 3.4 (that the transmission of a spike train through a synapse behaves like a linear system) in assessing the behavior of simple neural networks.

Example 3.5.1. Let us assume that the response curve of an excitatory synapse looks like an exponential decaying function activated after a delay. Such a response is shown in Figure 3.5.1. Let us now examine how the transmission through two such synapses in tandem would look.

Answer. To simplify our discussion, we describe each synapse as composed of two "boxes": a delay, followed by an exponential part. The two synapses in tandem will look like four boxes in tandem: a delay, an exponential, another delay, and another exponential. This situation is shown schematically in Figure 3.5.2A. In linear systems, the order through which the signal passes makes no difference. Therefore, we can

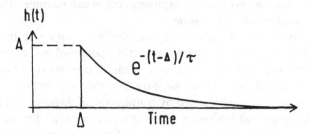

Figure 3.5.1. The shape of an "idealized" curve of excitatory postsynaptic potential.

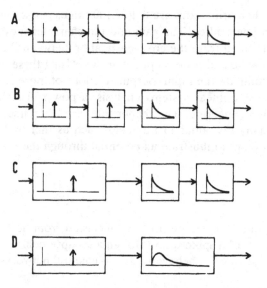

Figure 3.5.2. Transmission through two synapses in tandem. A: Decomposition of the transmission curves to a delay followed by an exponential. B: Changing the order of the operations: delays followed by exponentials. C: The two delays are merged into one delay twice as long. D: The two exponentials are merged.

rearrange the boxes so that the two delays are followed by the two exponentials (Figure 3.5.2B).

We can now merge the two delays into one box, which will give us a longer delay (the sum of the two delays). This situation is shown in Figure 3.5.2C. The last step in the process is to merge the two exponentials. For this we must convolve the two exponentials. In order to describe an exponential that extends only for positive time (e^{-kt} is defined

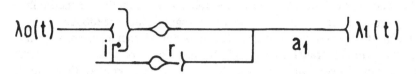

Figure 3.5.3. Recurrent inhibition.

for positive and negative t), we introduce the step function $U(t)$, which has a zero value from $-\infty$ to 0, and the value of unity from 0 to $+\infty$. The convolution of the two exponentials will read

$$h_{12}(t) = [U(t)Ae^{-kt}] * [U(t)Ae^{-kt}] = \int_{-\infty}^{\infty} U(\tau)Ae^{-k\tau} U(t-\tau)Ae^{-k(t-\tau)}d(\tau) \quad (3.5.1)$$

We can factor out the two constant amplitudes (A) and eliminate the step functions [$U(t)$] by changing the limits of integration to the range where both step functions are not zero.

$$h_{12}(t) = A^2 \int_{0}^{t} e^{-k\tau}e^{k(\tau-t)}d\tau \quad (3.5.2)$$

By solving the integral we arrive at

$$h_{12}(t) = A^2te^{-kt} \quad (3.5.3)$$

whose form is shown in the right-hand box in Figure 3.5.2D.

Exercise 3.5.1. How would the transmission through ten such synapses in tandem appear?

Exercise 3.5.2. What is the time delay until the peak of the response of ten such synapses in tandem is reached?

Example 3.5.2. Recurrent inhibition: Let us assume that we are facing a system in which a collateral of the postsynaptic axon branches off the main axon and excites an inhibitory interneuron, which in turn inhibits the first neuron. In neurophysiological jargon, this arrangement (shown in Figure 3.5.3) is called recurrent inhibition. Let us suppose that the cell is excited by an input spike train that has a rate function $\lambda_0(t)$. How will the output rate function $\lambda_1(t)$ appear?

Let $e(t)$ denote the transmission function of the synapse between the input axon (a_0) and the output cell (a_1), $r(t)$ the transmission function of the recurrent collateral, and $i(t)$ the transmission function of the inhibitory synapse. This network looks like a regular linear feedback system. It is most convenient to consider the behavior of such a system in terms of the Laplace transforms of all the functions, because the Laplace transform of a convolution of two functions is the product of their transforms. Let us therefore take the Laplace transforms of all the involved functions:

$\Lambda_0(s)$ – the input spike stream
$\Lambda_1(s)$ – the output spike stream
$Z(s)$ – the activity in the inhibitory interneuron
$E(s)$ – the transfer function between the input axon a_0 and the output axon a_1
$R(s)$ – the transfer function of the synapse between the recurrent collateral and the inhibitory interneuron
$I(s)$ – the transfer function of the inhibitory synapse

The main cell sums the excitatory and inhibitory effects:

$$\Lambda_1(s) = \Lambda_0(s)E(s) + Z(s)I(s) \qquad (3.5.4)$$

The activity in the interneuron (Z) depends only on the firing of the output cell:

$$Z(s) = \Lambda_1(s)R(s) \qquad (3.5.5)$$

From equations (3.5.4) and (3.5.5) we can derive the overall behavior of the network:

$$\Lambda_1(s) = \Lambda_0(s)(E(s)/[1 - R(s)I(s)]) \qquad (3.5.6)$$

This is similar to the conventional equation for a negative feedback loop. Much of its behavior can be inferred from the values of s for which the denominator $[1 - R(s)I(s)]$ becomes zero. For instance, if the real part of any of these s values is positive, the system is unstable.

With these examples we conclude this chapter on synaptic transmission. It should be clear to the reader that measuring the transmission curve of a synapse through the cross-correlation technique is a useful way of describing its properties, for it allows for integrating this synapse into larger networks and predicting their behavior.

The linear property that makes these calculations so simple is limited to firing rates for which the transmission functions do not vary. This limitation can be alleviated by measuring the transmission function of the synapse for every background firing rate separately and using it as the (variable) impulse response function in all the preceding formulas.

3.6 References

Abeles M. (1982a). Quantification, smoothing, and confidence limits for single-units' histograms. *J. Neurosci. Meth.* 5:317–25.

(1982b). *Studies of Brain Function. Vol. 6: Local Cortical Circuits: An Electrophysiological Study,* pp. 20–32. Springer-Verlag, Berlin.

Aertsen A. M. H. J. and Gerstein G. L. (1985). Evaluation of neuronal connectivity: Sensitivity of cross-correlation. *Brain Res.* 340:341–54.

Aertsen A. M. H. J., Gerstein G. L., Habib M. K., and Palm G. (1989). Dynamics of neuronal firing correlation: Modulation of "effective connectivity." *J. Neurophysiol.* 61:900–17.

Braitenberg V. (1978a). Cell assemblies in the cerebral cortex. In Heim R. and Palm G. (eds.), *Lecture Notes in Biomathematics, Vol. 21: Theoretical Approaches to Complex Systems,* pp. 171–88. Springer-Verlag, Berlin.

(1978b). Cortical architectonics: General and areal. In Brazier M. A. B. and Petche M. (eds.), *Architectonics of the Cerebral Cortex,* pp. 443–65. Raven Press, New York.

Cox D. R. (1962). *Renewal Theory.* Methuen, London.

Cox D. R. and Isham V. (1980). *Point Processes.* Chapman & Hall, London.

Dickson J. W. and Gerstein G. L. (1974). Interactions between neurons in auditory cortex of the cat. *J. Neurophysiol.* 37:1239–61.

Gerstein G. L. and Kiang N. Y.-S (1960). An approach to the quantitative analysis of electrophysiological data from single neurons. *Biophys. J.* 1:15–28.

Levick W. R., Cleland B. G., and Dubin M. W. (1972). Lateral geniculate neurons of cat: Retinal inputs and physiology. *Invest. Ophthalmol.* 11:302–11.

Michalsky M., Gerstein G. L., Czakowska J., and Tarnecki R. (1983). Interactions between cat striate cortex neurons. *Exp. Brain Res.* 51:97–107.

Palm G., Aertsen A. M. H. J. and Gerstein G. L. (1988). On the significance of correlations among neuronal spike trains. *Biol. Cybern.* 59:1–11.

Perkel D. H., Gerstein G. L., and Moore G. P. (1967). Neuronal spike trains and stochastic point processes. II. Simultaneous spike trains. *Biophys. J.* 7:419–40.

Siebert W. M. (1986). *Circuits, Signals and Systems.* McGraw-Hill, New York.

Toyama K., Kimura M., and Tanaka K. (1981). Cross-correlation analysis of interneuronal connectivity in cat visual cortex. *J. Neurophysiol.* 46:191–201.

4

Relations between membrane potential and the synaptic response curve

In Chapter 3 we discussed the input–output relations of synaptic connections, as measured by spike-train analysis. In this chapter we attempt to relate these phenomenological transmission curves to intracellular synaptic mechanisms. There appears to be no uniquely accurate way of translating the membrane potential changes into firing rates, though two extreme cases can be understood fairly well. The first case is that of a neuron whose membrane potential hyperpolarizes strongly after each action potential, and then begins to depolarize gradually until it hits the threshold and fires again. Firing times of such neurons are quasiperiodic. The second case is that of a neuron whose membrane potential fluctuates strongly around a constant mean level and whose excitatory postsynaptic potentials (EPSPs) and inhibitory postsynaptic potentials (IPSPs) are small.

Both types of neurons can be found in the mammalian nervous system. We refer to them as the periodically firing neurons and the randomly firing neurons. Sections 4.1 and 4.2 deal with the analysis of firing times in these two types of neurons. Section 4.3 describes the autocorrelation function and shows how it can be used to distinguish between the two types of neurons. It also shows that cortical neurons behave like randomly firing neurons. Sections 4.4 and 4.5 describe the expected modulations of firing rates generated by a postsynaptic potential in these two types of neurons.

4.1 Periodically firing neurons

Most of our knowledge of the electrophysiological behavior of mammalian nerve cells is derived from the classical work of Sr J. C. Eccles [1957] on the motoneurons of the spinal cord. In these neurons, activity in an excitatory synapse causes a brief depolarizing current. The voltage change caused by this current then passively decays back to the resting

118

level. Some of the inhibitory synapses produce the opposite effect by generating hyperpolarizing currents, and other inhibitory synapses increase the membrane conductance to chloride ions, a process that tends to clamp the membrane potential at a level below the threshold.

Some of the excitatory and inhibitory synapses act directly on the cell body, but the majority act on the dendrites. The local depolarizations and hyperpolarizations of the dendrites generate currents through the dendritic tree, some of which reach the cell body and affect its membrane potential. The synaptic potentials and currents flowing along the dendrites cause potential gradients there; however, the cell body is essentially equipotential.

The lowest firing threshold of the motoneuron is either at the region of emanance of the axon (the axon hillock) or at the initial segment of the axon. When the membrane potential of the cell body drives the potential of this sensitive region of threshold, the motoneuron fires. Each action potential is followed by a very large hyperpolarizing wave, which carries the membrane potentials far away from threshold. Following this hyperpolarization, the incoming EPSPs start a depolarizing trend that builds up until the membrane potential hits threshold and the neuron fires again. Therefore, in the motoneurons of the spinal cord, each action potential is followed by a prolonged period in which excitability is depressed, and the cells fire in quasi-periodic form. Many models of this behavior have been constructed [Rall, 1962; Stein, 1965; Calvin and Stevens, 1968; Holden, 1976; Sampath and Sirinivasan, 1977; Cope and Tuckwell, 1979; Tuckwell and Cope, 1980; Lansky, 1984].

Here we follow the description of the process as given by Calvin and Stevens [1968]. While the motoneuron fires at a steady rate, the after-potentials appear very similar, as can be seen by superimposing several recorded traces on top of each other (Figure 4.1.1). The late part of the after-potential looks like a ramp approaching the (fixed) threshold with a steady slope. Its points of intersection determines the average interspike interval. The depolarization (as it approaches the threshold) is driven by a multitude of synaptic potentials that bombard the huge dendritic tree of the motoneuron. These synaptic potentials arrive in random fashion and generate a fluctuating depolarization wave. The result is some "jitter" of the exact time at which the membrane potential hits threshold again. Calvin and Stevens were able to show that most of the "noise" in the interspike intervals could be attributed to synaptic noise.

In some motoneurons, the threshold is not fixed, but rather varies

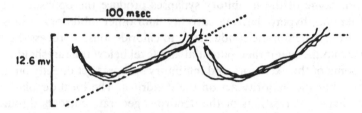

Figure 4.1.1. Intracellular membrane potentials in a motoneuron: superimposition of five after-potentials. The horizontal dotted line represents the threshold. The oblique dotted line represents the average rate of depolarization with which the membrane potential approaches threshold. The neuron fires in a quasi-periodic manner, with an average period of 118 ms. [From W. H. Calvin and C. F. Stevens, "Synaptic Noise and Other Sources of Randomness in Motoneuron Interspike Intervals," *Journal of Neurophysiology*, 31:574–87, 1968, with permission of The American Physiological Society.]

slightly during the interspike interval because of accommodation. This property generates a skewed distribution of interspike intervals. (See Calvin [1974] for measurements of threshold variations and their effects on the firing times of motoneurons.)

To summarize, neurons that exhibit large after-hyperpolarizations tend to fire periodically. The average period is inversely related to the strength of the excitatory drive to the neuron. The jitter around the mean period usually is small and is attributable to synaptic noise.

4.2 Randomly firing neurons

The cortical neurons are characterized as having a low spontaneous firing rate. The average firing rate is approximately five spikes per second, but about half the cells fire at less than two spikes per second. The firing pattern of these cells is highly irregular. Except for a brief initial refractoriness and a slight tendency to fire again after recovery from this refractory period, the autocorrelation curves (called also renewal density or conditional intensity curves) of these neurons are flat. In the following section we examine why this is so.

Let us assume that we have a very large network of neurons, some of which are excitatory and some of which are inhibitory. The cells are connected among themselves extensively, and they all fire "spontaneously" at a very low rate. This means that the membrane potential is below threshold most of the time. If we insert a micropipette into the

A   B

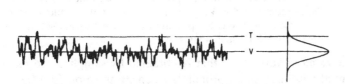

Figure 4.2.1. Fluctuations of the transmembrane potential. A: Sum of
100,000 simulated synaptic potentials per second arriving at random times. B:
The p.d.f. of the potentials of part A lying on its side. [Based on M. Abeles,
*Studies of Brain Function. Vol. 6: Local Cortical Circuits: An
Electrophysiological Study,* 1982, with permission of Springer-Verlag.]

cell body of such a neuron and record its membrane potential, we see
that it fluctuates in a seemingly random fashion around a constant mean.
If the neuron cannot fire an action potential, we expect to obtain a
record similar to that shown in Figure 4.2.1A.

4.2.1 The membrane potential as a random variable

We now wish to treat the membrane potential as a random
variable and to relate its statistics to the incoming EPSPs and IPSPs. In
the following paragraph we go through the mathematical formulation
required to do so. This paragraph is presented in order to assist those
interested in finding relevant examples for teaching probability theory to
students of biology. The reader with no background in probability
theory may find it complicating and may instead choose to skip ahead
and use intuition regarding the average and variance of a noisy waveform.

First we must define a probability space $(\Omega, \mathbf{B}, \mathbf{P})$ in which our measurement takes place. We do this in the following manner:

Ω (the sampling space) is the time interval that spans $(-T, T)$, where T can
become as large as we wish.

$\mathbf{B}$ is the field of all the events for which we wish to assess the probability of
occurrence. It contains all the points in time $(t, t \in \Omega)$, all the continuous time sections $[t_1, t_2]$ where $-T \le t_1 < t_2 < T$, and all the
subsets that can be formed from these times by a countable number
of unions and intersections, or by complementing.

$\mathbf{P}$ is the probability of obtaining any event defined in $\mathbf{B}$. We assume that all
possible time points t_0 may be selected with the same probability and
that the probability of obtaining a time within $[t_1, t_2)$ is proportional
to $t_2 - t_1$ (which also defines the probability of all other events in $\mathbf{B}$).

The random variable V is defined as follows: If we select a point in time t, the value of the membrane potential at that time $v(t)$ is the value of the random variable V, and the probability of obtaining any particular value of $V = v$ is the probability of selecting the set of all the time points for which $V(t) = v$.

The random variable V that is obtained by sampling the membrane potential will have some probability density function (p.d.f.), as illustrated in Figure 4.2.1B, with an average μ and variance σ^2. The average and the variance may be determined by averaging the membrane potential and the squared membrane potential over time in the following way:

$$\mu = \lim_{T \to \infty} \left\{ (1/2T) \int_{-T}^{T} v(t)\, dt \right\} \qquad (4.2.1)$$

$$\sigma^2 = \lim_{T \to \infty} \left\{ (1/2T) \int_{-T}^{T} [v(t) - \mu]^2\, dt \right\} = \lim_{T \to \infty} \left\{ (1/2T) \int_{-T}^{T} v^2(t)\, dt - \mu^2 \right\} \qquad (4.2.2)$$

Whenever we need numerical values for these parameters, we use $\sigma = 5$ mV and $\mu = -60$ mV.

If the membrane potential is generated only by postsynaptic currents, if all these synaptic potentials appear at independent times, and if the synaptic potentials are added linearly, then we can expect the p.d.f. of the membrane potential to be normal; that is,

$$f_V(v) = \exp\{-(v - \mu)^2/2\sigma^2\}/(2\pi\sigma^2)^{1/2} \qquad (4.2.3)$$

None of the foregoing assumptions is completely accurate. The fluctuation of the membrane potential might also have some metabolic components, the firing times of different neurons show some degree of (weak) correlation, and the synaptic summation is not linear. Therefore, p.d.f. values for the membrane potentials of real cortical neurons are found to deviate somewhat from the normal shape [Ellul, 1972].

In the following discussion we assume that the membrane potential is distributed normally. This assumption facilitates calculation of the probabilities of events. Such calculations will give us qualitative insight into the behavior of neurons in a large network. The reader should keep in mind that the results we obtain are only rough approximations.

Let us assume that the membrane potential fluctuations are generated by the summation of many postsynaptic potentials that has an exponential form:

$$s(t) = Ae^{-t/k} \quad \text{for } t \geq 0 \quad \text{and} \quad s(t) = 0 \quad \text{for } t < 0$$

Let us further assume that all the postsynaptic potentials have the same amplitude A and the same time constant k. If these synaptic potentials arrive at an average rate of λ per second in a Poissonian way (i.e., for every brief time interval Δt, a synaptic potential will occur with a probability $\lambda \Delta t$, irrespective of what happened in the past), then the average depolarization (hyperpolarization) and variance of the fluctuations of the membrane potential are given by

$$\mu = \lambda \int_0^\infty A e^{-t/k} dt = \lambda A k \qquad (4.2.4)$$

$$\sigma^2 = \lambda \int_0^\infty (A e^{-t/k})^2 dt = \lambda A^2 k/2 \qquad (4.2.5)$$

See Rice [1954] for detailed derivations of these relations.

Exercise 4.2.1. How will the average and the variance appear if there are n different independent streams of synaptic potentials, where the ith potential looks like

$$s_i(t) = A e^{-t/k_i},$$

and it occurs λ_i times per second? Note that if the IPSP looks exactly like the EPSP, except for the inversion in direction (i.e., $A_{EPSP} = -A_{IPSP}$), then equation (4.2.5) remains valid.

Example 4.2.1. The "average" cortical cell receives 20,000 synapses from different neurons. Let us assume that each of the presynaptic cells fires at a rate of five spikes per second. Then the rate of arrival of postsynaptic potentials is $\lambda = 100,000$ per second. While measuring the membrane potential of such a cortical cell, it was observed that the variance of its fluctuations is 25 mV^2, and the time constant of the decay of the PSPs is 10 ms. What is the average amplitude of a synaptic potential?

If we substitute the measured values into equation (4.2.5) and solve for the amplitude, we obtain

$$25 = 100,000 A^2 \cdot 0.01/2$$
$$A^2 = 0.05 \text{ mV}^2$$
$$A = 0.22 \text{ mV}$$

This is somewhat higher than what was actually recorded in cortical slices [Komatsu et al., 1988].

Exercise 4.2.2. How will our evaluation of A in Example 4.2.1 differ if we do not assume that all the cortical cells fire at a rate of five spikes per second, but that five per second is merely the average firing rate, around which there is some scatter?

Exercise 4.2.3. How will our evaluation of A in Example 4.2.1 differ if we do not assume that all the PSPs have the same time constant, but that the average time constant is 10 ms?

Exercise 4.2.4. If the amplitudes of the various PSPs are not equal, what can we deduce from the variance of the membrane fluctuations?

4.2.2 The autocorrelation function

In this section we discuss ways in which the properties of the postsynaptic potentials affect the characteristics of the membrane potential fluctuations. The major intent is to show how we can extract some parameters of the synaptic potentials from the autocorrelation function of the membrane potential. The reader who is not interested in this topic may skip this section and go directly to Section 4.2.3.

In the discussion that follows, we assume that the membrane potential is normalized so as to have a zero mean and a variance of unity; that is,

$$u(t) = [v(t) - \mu]/\sigma \qquad (4.2.6)$$

Throughout the following subsections we refer to the normalized membrane potential as $u(t)$, and to the unnormalized (measured in mV) as $v(t)$.

If we measure the membrane potential at two times separated by a fixed delay of τ, we obtain two random variables $u(t)$ and $u(t + \tau)$.

This process can be accurately defined as follows: Let us examine the probability space $(\Omega, \mathbf{B}, \mathbf{P})$, in which we sample time with equal probabilities of sampling for all time instants, as described in Section 4.2.1. In this space we perform the following trial: We select a random time instant t. For every selection, we obtain two numbers: the value of the membrane potential at time t $[u(t)]$ and the value of the membrane potential at $t + \tau$ $[u(t + \tau)]$. Note that we need only one trial (in which we obtain t) to obtain a pair of random variables.

Each of these random variables has a mean of zero and a variance of unity, but they are not necessarily independent random variables. For instance, if the delay τ is extremely short, the value of the membrane potential at $t + \tau$ will be very close to its value at t. The degree of independence of the two measurements can be derived from their correlation coefficient. The correlation coefficient r is defined by

$$r(\tau) = \lim_{T \to \infty} \left\{ (1/2T) \int_{-T}^{T} u(t)u(t + \tau)\, dt \right\} \qquad (4.2.7)$$

For each time delay τ we can obtain another correlation coefficient. In other words, the value of the correlation coefficient r is a function of the delay τ. This function is called the autocorrelation function $[r(\tau)]$. The autocorrelation function must by symmetric:

$$r(\tau) = r(-\tau)$$

If the membrane potential looks like a Gaussian process, then the two random variables $x = u(t)$ and $y = u(t + \tau)$ behave like a two-dimensional normal random variable. Their two-dimensional probability density function is given by

$$f_{X,Y} = \exp\{(x^2 - 2rxy + y^2)/[2(1 - r)^2]\}/[4\pi(1 - r)^2]^{1/2} \qquad (4.2.8)$$

In this joint probability density function, r is the correlation coefficient, which depends on the delay (τ) between the two samples x and y.

Exercise 4.2.5. How will the autocorrelation function of a periodic waveform appear?

Exercise 4.2.6. How will the autocorrelation function of a "white noise" appear?

Exercise 4.2.7. Let us construct a random telegraph wave in the following manner: We begin with the value of 1. The wave remains at this value until time T, when we toss a coin. If the coin falls 'head up,' we flip the value of the wave to -1. If the coin falls "tail up," we leave the waveform at 1. If we continue this process at $2T$, $3T$, $4T$, and so forth, the result is a waveform that has the value of either 1 or -1 and has a probability of $\frac{1}{2}$ of reversing its polarity at times T, $2T$, $3T$, and so forth. How will the autocorrelation function of this random telegraph wave look?

One of the most important features of the autocorrelation function is that its Fourier transform is equal to the average power spectrum of the original signal. This relation is stated as follows:

$$\int_{-\infty}^{\infty} r(t)e^{-i\omega t}dt = \lim_{T\to\infty}\left\{ (1/2T)\cdot\left|\int_{-T}^{T} u(t)e^{-i\omega t}dt\right|^2 \right\} \qquad (4.2.9)$$

Another important feature of the autocorrelation function is that the autocorrelation of a sum of two uncorrelated functions is the sum of their autocorrelations. Expressed in formula form, this statement reads as follows: Let $u(t)$ and $w(t)$ be two normalized (zero mean and unit variance) waveforms. If for every τ we have

$$\lim_{T\to\infty}\left\{ (1/2T)\cdot\int_{-T}^{T} u(t)w(t + \tau)\, dt \right\} = 0$$

[i.e., $u(t)$ and $w(t)$ are uncorrelated], then if we denote their sum by $x(t)$, we have

$$r_x(\tau) = r_u(\tau) + r_w(\tau) \qquad (4.2.10)$$

In general, let us assume that we have a complex waveform composed of many (N) primitive waveforms. Let us further assume that these waveforms are pairwise uncorrelated. Then the autocorrelation of this composite waveform is the sum of the autocorrelations of the primitive waveforms, and its power spectrum is the sum of the power spectra of the primitive waveforms.

This property is used extensively to analyze the characteristics of ionic channels in membranes (so-called noise analysis). In our case we can use it to analyze the membrane potential fluctuations. If these fluctuations consist of many uncorrelated postsynaptic potentials, then the autocorrelation and the power spectrum of the membrane potential should look like the sum of the autocorrelations and the sum of the power spectra of the individual synaptic potentials.

Exercise 4.2.8. A pulse function $p(t)$ is any function for which the area under the curve of the product of the function with itself is bounded:

$$\int_{-\infty}^{\infty} p(t)p(t + \tau)\, dt < \infty$$

1. What is the autocorrelation function of $p(t)$?
2. The pulse $p(t)$ appears λ times per second at random times. What is the autocorrelation of this waveform?

If we assume that all the synaptic potentials have the form $Ae^{-t/k}$, that they appear at random times, and that they add up linearly, then the autocorrelation of the membrane potential is

$$r(\tau) = \lambda A^2 \int_0^{\infty} e^{-t/k} e^{-(t+\tau)/k} \, dt$$

For positive delays we have

$$r(\tau) = \lambda A^2 \int_0^{\infty} e^{-(2t+\tau)/k} \, dt = \lambda A^2 k [e^{-(2t+\tau)/k}]_0^{\infty}/2$$

$$r(\tau) = (\lambda k A^2/2) e^{-\tau/k} \qquad\qquad (4.2.11)$$

This autocorrelation function will have the same time constant as the elementary postsynaptic potentials. Therefore, it may be used to evaluate the time constant of the synaptic potentials.

Note that in the computation of equation (4.2.11) we did not normalize the membrane potential (to have zero mean and unit variance), and therefore what we obtained was not the proper autocorrelation function, as defined in equation (4.2.7). When $f(t)$ has a zero mean, the proper name for this function is the covariance function; however, it is customary to call any function of the form

$$r(\tau) = \lim_{T \to \infty} \left\{ (1/2T) \cdot \int_{-T}^{T} f(t) f(t + \tau) \, dt \right\}$$

an autocorrelation function.

With this we conclude our discussion of the relations between the postsynaptic potentials and the membrane potentials. The reader is referred to the work of Rice [1954] for a thorough discussion of noise analysis.

4.2.3 Firing rate and membrane potential

In the preceding sections we assumed that the neuron was unable to fire action potentials. However, one of the important purposes of studying the statistics of the fluctuations in the membrane potential is to relate it to the firing rate of the neuron.

As a starting point, let us assume that the cell has a fixed threshold at the level of θ above the resting (average) membrane potential. Let us further assume that whenever the membrane potential rises above this level (θ), the cell fires an action potential and becomes refractory (unresponsive) for r ms. After this time, if the membrane potential is still above threshold (θ), the cell will fire again. If, when the cell has recovered from its refractory period, its membrane potential is below θ, it will fire again only when the membrane potential reaches the threshold again.

These assumptions ignore the relative refractory period and the effects that slow changes in the membrane potential have on the threshold. This model is very different from the model of the periodically firing neuron presented in Section 4.1, as here we do not assume that a marked and prolonged hyperpolarization follows each action potential, nor do we assume that the action potential resets all the synaptic currents that were initiated before its occurrence. Although these assumptions may appear strange, we shall show that they are appropriate for most randomly firing neurons.

More comprehensive models for the statistics of firing times are available in the literature. They are commonly known as models for the first passage time. In these models, the membrane potential is reset to an initial value v_0 after each spike. Then the streams of EPSPs and IPSPs change the membrane potential until it hits the threshold θ and the neuron fires again. The statistics of the time required to reach θ when starting from v_0 are identical with those of the interspike time interval. The reader is referred to the work of Lansky [1984] for a review and mathematical analysis of such models of the first passage time. In the following discussion we describe a much simplified model that relates the firing rate to the mean and variance of the fluctuations of the membrane potentials.

In Section 4.2.1 we treated the membrane potential as a random variable whose values are obtained from sampling the potential at randomly selected time instants. Here it is more useful to examine the values of the membrane potentials of many similar cells at a given moment t. The range of possible values obtained is the range of membrane potentials that our set of similar cells may have at time t. The probability of finding a certain value v is the probability of selecting those cells in which the membrane potential happens to be v at time t. This is the approach used when analyzing stochastic processes, which can be formalized as follows:

Let $(\Omega, \mathbf{B}, \mathbf{P})$ be a probability space in which the sample space (Ω) is a set of cortical neurons with similar properties. The basic "experiment" is to randomly sample one of the cells from Ω. All cells in Ω have the same probability of being selected. The events (members of $\mathbf{B}$) whose probabilities are given by the function $\mathbf{P}$ are subsets of cells (from Ω). $\mathbf{B}$ is the set of all such events. The rule that all cells have the same probability of being sampled defines a probability for every possible event in $\mathbf{B}$. To each cell that we sample we attach a function of time $[v_i(t)]$: the membrane potential of that cell. Obtaining this ensemble of membrane potentials in this fashion, with the probabilities of obtaining them, is called a stochastic process. For every time instant t, the set of values attained by these membrane potentials $\{v_i(t)\}$ is a random variable.

The random variable $v(t)$ (i.e., the membrane potential at time t) has an expected value $E[v(t)]$, a variance $E[v^2(t)] - E[v(t)]^2$, a probability density function $f_V(v, t)$, and a probability distribution function $F_V(v, t)$. The latter gives us (at time t) the probability of obtaining a membrane potential smaller than or equal to v for every value v. We note in particular that the probability of the membrane potential being above threshold (θ) at time t is

$$\mathrm{pr}\{v(t) > \theta\} = 1 - F_V(\theta, t) \qquad (4.2.12)$$

Note that the statistics described here are not identical with those obtained by averaging over time, as in Section 4.2.1. Here, we obtain a random variable by sampling the membrane potentials of many cells at one time instant. In Section 4.2.1 we obtained a random variable by sampling the membrane potential of one neuron at many time instants. In general, the two statistics are not necessarily identical. However, for stochastic processes that are *ergodic*, the properties determined by sampling over time are almost always (i.e., with a probability of 1) equal to the properties determined by sampling across the ensemble of functions.

In reality, it is impractical to sample a great many neurons simultaneously. Therefore, if the properties determined by sampling one neuron are stationary (do not vary over time), we can assume that what we have measured is a realization of an ergodic stochastic process. Then we can use the time-averaged properties as those of the process. However, mathematical considerations are facilitated by considering stochastic processes and by averaging across the ensemble of functions, as will be done later.

If for a given neuron the membrane potential is above threshold at time t, then the cell either starts firing at t or is refractory (i.e., it has fired within the last r ms). If, however, the membrane potential is below θ, then the cell does not start firing at this instant, but may have started to fire in the last r ms.

Suppose we want to measure the firing rate of our group of cells at time t by observing the time interval $(t - r, t)$ and counting how many cells fire during that time. If we observe N cells, and n of them fire, we say that

$$\lambda(t) \approx n/(Nr) \qquad (4.2.13)$$

This is so because we measure the activity from each of the N neurons during r ms, so that Nr is the total time of measurement. During this time we observe n action potentials. Therefore, the firing rate is n/Nr. As N becomes larger and larger (as we sample more and more cells at time t), the ratio n/N is the probability that a cell will fire somewhere between $t - r$ and t. This probability is somewhat greater than that of the cell having a membrane potential above θ, at time t, because the membrane potential may be just above threshold at time $t - r$ and decline below it at time t. Therefore, we can write

$$\lambda(t)r \approx \lim_{N \to \infty}(n/N) \geq \text{pr}\{v(t) > \theta\} \qquad (4.2.14)$$

Note that when the fluctuations of the membrane potential are slow, relative to r, the difference between n/N and the probability is small.

The probability on the right side of equation (4.2.14) is easily derived from the statistics of the random variable $v(t)$, as we shall see shortly. The foregoing argument suggests that the probability that the membrane potential is above threshold at time t gives us a lower boundary on the firing rate around this time. On the other hand, when we derived equation (4.2.14) we considered only the absolute refractory period (r). Were we also to take into account the relative refractory period, we would have cases in which the membrane potential would be above threshold during both $(t - r, t)$ and $(t - 2r, t - r)$, with the neuron firing only once. If that happened frequently (as would be the case if the neuron fired at a high rate), then the fraction of instances at which the neuron would fire (n/N) would be smaller than the probability of the membrane potential being above threshold. Both effects are small at low firing rates, and they tend to cancel each other out. In the future discussion, we shall use the formula

$$\lambda(t) = \text{pr}\{v(t) > \theta\}/r \tag{4.2.15}$$

Notice that the firing rate λ is not necessarily constant, but may vary over time. For instance, if t is the time elapsed since a stimulus was given, $\lambda(t)$ is the expected response of the group of cells (Ω) to that stimulus.

Equation (4.2.15) seems peculiar because when r (the refractory period) becomes very small, λ (the firing rate) becomes great. This is justified in our model because the rate of fluctuations of the membrane potential $[v(t)]$ is limited. Therefore, once the membrane potential is above threshold, it will remain there for a while. Had the refractory period been much smaller than this period, the neuron would have fired many times during every period for which $v(t) > \theta$.

Problem 4.2.1. For simplicity, let us assume that the membrane potential looks like a low-pass Gaussian noise. That is, $v(t)$ has a normal probability density function, at two time instants t_1 and t_2 the two random variables $v(t_1)$ and $v(t_2)$ have a joint Gaussian distribution [such as equation (4.2.8)], and the autocorrelation function of the membrane potentials decays to zero exponentially. Derive the probability of having a spike somewhere between $t - r$ and t [i.e., find an equation similar to (4.2.14), but without the > sign]. The reader can consult Chapter 3.6 of Cox and Isham [1980] for treatment of similar problems.

Problem 4.2.2. Investigate equation (4.2.15) by simulations, and find the range of parameters and assumptions for which it holds true.

In the discussion thus far, we have ignored two additional factors that affect the firing rate of a cell: the relative refractory period, and the effect that an action potential has on the membrane potential after its termination. We can take the relative refractory period into consideration by letting the threshold (θ) vary, so that after the end of the absolute refractory period it will be high, and will drop gradually to the resting level.

The effect of the action potential on the membrane potential depends on the structure of the nerve cell and on the distribution of the synaptic inputs to the cell. In the cell body, the repolarizing currents at the tail of the action potential are huge compared with the synaptic currents. Therefore, all the synaptic effects caused by currents that end before the action potential are "forgotten." We say that the action potential "resets" the membrane potential at the soma. This resetting effect is carried also for some unknown distance into the large dendrites. However, the dendrites that are far away from the soma are affected very weakly by the action potential at the soma. If the remote dendrites are depolarized before the action potential, they will continue to withdraw current from the soma after the action potential. This situation is likely to be even more pronounced when the dendritic spines are considered. There the narrow stalks, which have high ohmic resistance but relatively large shunting capacitance, will poorly transmit the echo of the brief action potential. If the head of the spine is depolarized by an EPSP prior to the firing in the soma, it probably will remain depolarized and continue to withdraw current after the action potential. In the majority of the cortical cells, the soma and the proximal dendrites have very few, if any, excitatory synapses. Most the excitation of the pryamidal and spiny stellate cells is derived from the dendritic spines. Consequently, the resetting effect of the action potential on these cells is relatively small.

The action potential may continue to affect the membrane potential for a while by altering the ionic processes at the soma and large dendrites. The conductance of potassium is generally elevated after the action potential. This causes the soma to hyperpolarize and diminishes the effects of the synaptic currents on the membrane potential. In some cells, the action potential triggers an increase in calcium conductance that tends to depolarize the cell. If few action potentials are fired in succession, the sodium–potassium exchange pump may be turned on metabolically. This pump is often electrogenic (it pumps out more sodium ions than it pumps in potassium ions) and will hyperpolarize the cell. Action potentials may cause potassium to accumulate in the very narrow intercellular space around the cell body. Such potassium accumulation will depolarize the cell. The overall balance of these effects cannot be assessed without detailed knowledge of the morphology and all the electrical properties of the cell and its surrounding extracellular space. Quantitative consideration of these effects is beyond the scope of this book.

To summarize, the randomly firing neuron is characterized by a noisy membrane potential that fluctuates around a steady mean potential. To a rough approximation, its firing rate is proportional to the probability that the membrane potential is above threshold.

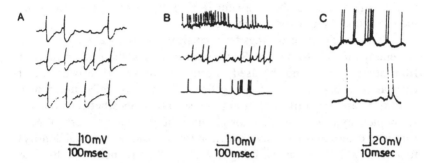

Figure 4.3.1. Intracellular recordings from neurons (A and B from the spinal cord). A: Recordings from neurons identified as motoneurons. B. Recordings from cells not identified as motoneurons. [From L. L. Glenn and W. C. Dement, "Membrane Potential, Synaptic Activity, and Excitability of Hindlimb Motoneurons during Wakefulness and Sleep," *Journal of Neurophysiology,* 46:839–54, 1981, with permission of The American Physiological Society.] C: Recordings from motor cortex of cat. [From C. D. Woody and P. Black-Cleworth, "Difference in Excitability of Cortical Neurons as a Function of Motor Projection in Conditioned Cats," *Journal of Neurophysiology,* 36:1104–16, 1973, with permission of The American Physiological Society.]

4.3 The autocorrelation function for spike trains

Sections 4.1 and 4.2 described two types of neuronal activity: periodic and random firing. The former mode of activity was attributed to neurons that show large after-hyperpolarizing potentials in addition to strong (net) excitatory drive. The random mode was attributed to neurons that do not show marked after-potentials and whose excitation is derived mainly from portions of the dendritic tree whose potentials are not reset by the action potential in the cell body. Do such neurons exist in the mammalian nervous system?

Figure 4.3.1 shows samples of intracellular recordings taken from spinal cord neurons of moving (unanesthetized) cats. Figure 4.3.1A shows recordings obtained from cells identified as motoneurons. The large hyperpolarizing wave that follows each action potential is clearly seen, as is the almost periodic firing pattern of the cell. Figure 4.3.1B shows similar recordings from cells not identified as motoneurons. Presumably, these are interneurons of the spinal cord. After-hyperpolarizations are not seen at all or are very small (middle trace), the firing times look random, and a "noisy" membrane potential can be observed (top trace). Figure 4.3.1C shows a similar recording from a neuron in the motor cortex of an unanesthetized cat.

Such intracellular recordings from cortical cells of awake animals are extremely difficult to obtain. Recordings from anesthetized animals or from brain slices cannot be used to estimate the statistical behavior of cortical cells because the ongoing firing patterns of cells under such conditions are very different from those seen in unanesthetized animals. A simple way to obtain some insight into the behavior of cortical neurons is to examine their autocorrelation (also known as renewal density) functions, which give us the firing rate of a cell as a function of the time elapsed since an earlier action potential. In Section 4.2.2 we described the time-averaged autocorrelation function for a continuous function $u(t)$ (the membrane potential), which was obtained by

$$r(\tau) = \lim_{T \to \infty} \left\{ (1/2T) \int_{-T}^{T} u(t) \cdot u(t + \tau) \, dt \right\} \qquad (4.3.1)$$

For a spike train, we can replace the continuous function $u(t)$ by a series of impulses $\Sigma\delta(t - t_i)$ placed at the firing times $\{t_i\}$ in the spike train, and run through the computation of equation (4.3.1). (See Perkel et al. [1967] for a more detailed description.)

The same function (except for a scaling factor) can be obtained in a much simpler way. First, consider each spike as a stimulus, and construct a poststimulus time histogram (PSTH) for the spikes that follow it. Then normalize the histogram to obtain a firing rate, as described in Section 3.1. What we obtain is a function that describes the expected firing rate at various delays after the cell fires. Thus, if the cell is firing in a periodic manner, the firing rate at delays of one, two, three, or more periods after the cell fires will be high again. That is, the autocorrelation function of a periodic spike train will also be periodic, with the same period as the spike train. If the firing pattern is completely random (Poissonian), then for any delay the expected firing rate is constant (λ) and equal to the average firing rate of the spike train.

In statistics, the function that describes the rate of occurrence of an event as a function of the time elapsed since its previous occurrence is known as an "intensity function." If the process generating the spike train is a renewal process (i.e., a process in which the probability of occurrence of a spike is a function only of the time elapsed since the last spike [Cox, 1962]), this function is also called the renewal density. How-

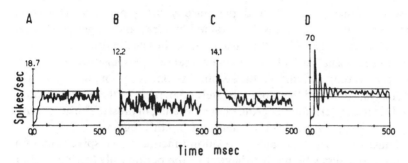

Figure 4.3.2. Autocorrelation functions for cortical cells. A–C: Cells from auditory cortex. D: Cell from secondary somatosensory cortex.

ever, physiologists are accustomed to using the term "autocorrelation function," and we shall therefore continue to use that name.

The autocorrelation functions of Figure 4.3.2 were computed from spike trains recorded extracellularly in unanesthetized animals. Most of the fluctuations seen are due to the limited amount of data from which the curves were computed. The two horizontal lines show the range in which we expect to find 99 percent of a stationary Poisson spike train with the same average firing rate.

The prolonged depression seen in Figure 4.3.2A indicates that in this neuron, each action potential is followed by a prolonged (about 70 ms) period during which excitability is depressed, probably because of the presence of after-hyperpolarization. The autocorrelation shows that the cell does not fire in any periodic form. This must mean that the total excitatory drive for this neuron is not strong enough to drive the average membrane potential above threshold. We might expect that if this type of neuron were driven to fire at rates above fifteen spikes per second, it would fire periodically, as does the motoneuron. However, in the auditory cortex in the cat, only 7 percent of the cells showed autocorrelations with such a shape. Most of the cells had only a brief period of low excitability, which was followed immediately by a return to the average firing level (as in Figure 4.3.2B) or was followed by a period of increased excitability (as in Figure 4.3.2C). Neurons with such autocorrelations behave like the randomly firing neuron described in Section 4.2.

In our laboratory we have made recordings and computed autocorrelation functions for more than 2,000 neurons in the auditory, frontal, prefrontal, and posterior parietal regions in awake monkeys. Only in a few cases have we observed periodically firing neurons. This is in marked contrast to the results obtained by Gray and Singer [1989] and Eckhorn et al. [1988] in the visual cortex in anesthetized cats, where prolonged visual stimulation evoked marked periodicities. However, when recording activities in the secondary somatosensory area in awake monkeys, we found that approximately 25 percent of the neurons exhibited clear-cut periodicities [Ahissar and Vaadia, in press]. The autocorrelation of Figure 4.3.2D is for a neuron from this region.

Thus, we can use the autocorrelation function of the spike train of a neuron to assess whether it behaves like the periodically firing model or like the randomly firing model. For large cortical areas in awake animals, the randomly firing model is adequate to describe the neuronal activity.

Problem 4.3.1. Build a model of a neuron whose firing times follow autocorrelation functions such as those in Figure 4.3.2, and show the relation between the following two variables: $\text{pr}\{v > T\}$, obtained when the cell is not allowed to fire, and λ, the average firing rate of the cell.

To summarize, the changes in the excitability of a neuron that are brought about by firing an action potential can be assessed by computing the autocorrelation function of the neuron's spike train. For most cortical regions it is safe to assume that the membrane potential fluctuates most of the time well below threshold. Occasionally, the potential hits the threshold, and then the cell fires. The brief refractory period that follows firing prevents another firing, but within a short time the excitability of the cell returns to its resting value.

4.4 Synaptic effects on periodically firing neurons

In this section we wish to analyze the effect that a single synaptic potential might have on the firing rate of the postsynaptic neuron for a periodically firing neuron (such as the motoneuron of the spinal cord). In these cells (as described in Section 4.1), each action potential is followed by a prolonged hyperpolarization that reduces the cell's excitability for several tens of milliseconds. According to Knox [1974] and Knox and Popple [1977], the firing rate of the motoneuron should follow the derivative of the postsynaptic potential. This means that for an EPSP, the cell should increase its firing rate during the rising phase and decrease its firing rate during the repolarizing phase. Qualitatively, one could argue that if the cell fired during the rising phase of the EPSP, it would not fire again for some time. If for some reason the large EPSP came at a moment when the cell was hyperpolarized and therefore did not trigger a spike during the rising phase, it would have a very small chance of firing when the membrane was repolarizing. These properties suggest that the response of the cell to the EPSP should look somewhat like the derivative of the EPSP. Indeed, Knox and associates were able to show that when a cell had only one presynaptic input and there was no additional source of membrane noise, the synaptic transmission curve looked like the derivative of the EPSP. Recently, Knox [1981] reported that the responses of a motoneuron to intracellular stimulation indeed followed the predictions derived from his equations.

Kirkwood [1979], working experimentally on motoneurons, found that the derivative rule was too extreme. He suggested that a sum of the EPSP itself and its derivative would give a reasonable approximation to the cross-correlation between a presynaptic fiber and a motoneuron. In more recent work, in which he studied the effects of weak synapses on motoneurons, he found that the shape of the response did not show a phase of depressed firing rate as was implied by the derivative rule [Kirkwood and Sears, 1982, Figure 1].

Fetz and Gustafsson [1983] carefully compared intracellularly recorded synaptic potentials with cross-correlations. In the remainder of this section, we follow their line of thought.

The regular sequence of affairs for the periodically firing neuron can conveniently be described in the following manner (Figure 4.4.1A): The average membrane potential that follows each spike is flattened and drawn on the abscissa. The distance between the threshold and the (flattened) membrane potential is shown as a falling line that hits the membrane potential at the time of the average interspike interval (compare Figure 4.4.1A and Figure 4.1.1).

If an EPSP (dashed line in Figure 4.4.1B) arrives too early, it will not elicit any postsynaptic spike. The earliest EPSP that might generate a spike is the one shown by the solid line in Figure 4.4.1B, and the latest EPSP that might generate a spike is shown by the dotted line. One can argue that the effect of these EPSPs is not exactly to add a spike to the postsynaptic spike train, but rather to shift slightly the time of appearance of the next spike.

In order to see how the EPSP is expressed in the cross-correlation, we should synchronize the graphs of Figure 4.4.1B to the presynaptic spike,

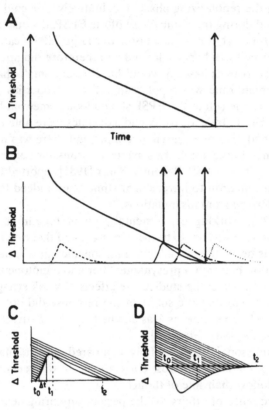

Figure 4.4.1. Effects of synaptic potentials on the firing times of a periodically firing neuron. A: The normal period. B: Shortened periods due to EPSP. C: Effect of an EPSP. D: Effect of an IPSP.

rather than to the postsynaptic spike (as was done in Figures 4.4.1A and B). This is done in Figure 4.4.1C. The presynaptic spike may occur at any point in time (relative to the previous postsynaptic spike). Therefore, what appeared as a single falling threshold line in Figure 4.4.1B may now appear anywhere along the time axis. This situation is represented in Figure 4.4.1C by falling lines, which are equally spaced over time. The time at which the EPSP may cause the postsynaptic cell to fire is (t_0, t_1), whereas during (t_1, t_2) the postsynapatic cell will never fire. This means that the cross-correlation between the spike trains of the presynaptic and postsynaptic neurons will show an elevation at (t_0, t_1) and a depression during (t_1, t_2). The probability of finding a spike during the time Δt is proportional to the number of threshold lines that cross the section of the EPSP that is above Δt.

For an IPSP (Figure 4.4.1D), we get a deep depression of firing rate during (t_0, t_1), with a mild elevation during (t_1, t_2). Note that the exact point at which t_1 occurs, for both EPSPs and IPSPs, depends on the slope of the threshold lines (i.e., on the firing frequency of the postsynaptic neuron).

If the membrane potential of the postsynaptic neuron is noisy (as it always is in reality), the threshold *lines* of Figure 4.4.1 should be replaced by *bands*. The points t_0, t_1, and t_2 become blurred by the noise. If the synaptic potential is small, relative to the background noise, the initial hump (or dip) is spread and occupies a larger portion of the postsynaptic potential.

An important experimental finding of Fetz and Gustafsson [1983] was that the onset of the hump in the cross-correlation was delayed by about 1 ms relative to the onset of the EPSP. This means that when one is cross-correlating the presynaptic spike train with the postsynaptic train, the expected delay is 1.5–2 ms.

If the firing rate becomes low (less than about five spikes per second for the motoneuron [Calvin, 1974]) and irregular, if the background synaptic noise increases and the individual postsynaptic potentials diminish, then we approach the case of the randomly firing neuron described in the next section.

Figure 4.4.2 shows autocorrelation functions for a neuron in the secondary somatosensory cortex of an awake monkey. While the neuron was active at a high rate, its firing pattern was periodic (Figure 4.4.2B), whereas at a lower firing rate it appeared to fire randomly (Figure 4.4.2A).

4.5 Synaptic effects on randomly firing neurons

We now turn again to the randomly firing neuron, described in Section 4.2, to examine the effect of a synaptic potential on its firing rate. This issue is dealt with separately for small postsynaptic potentials and for large postsynaptic potentials.

4.5.1 The effect of small synaptic potentials

Imagine the following hypothetical experiment: We are measuring the intracellular potentials of many (N) neurons simultaneously. All

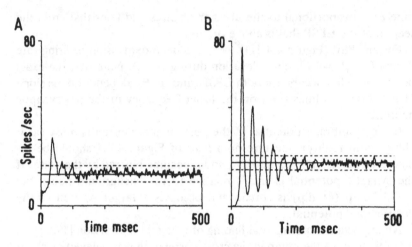

Figure 4.4.2. Switching from periodic firing to random firing. Autocorrelations for a neuron in secondary somatosensory cortex. [Courtesy of A. Ahissar and E. Vaadia.]

these cells are of the same kind and have the same statistics. We also assume that by means of pharmacological treatment we have managed to prevent the firing of these neurons, but not of the cells that give input to the measured neurons. Thus, we measure all the membrane potential fluctuations that arise from the incoming EPSPs and IPSPs, but we do not see the large voltage swings that are actively generated by the membrane during the action potential. At a given moment t, we sample the voltages of all these cells. The values we obtain $\{u(t)\}$ are the values of a random variable (u). Let us assume that the units of the membrane potential are normalized so that its expected value is zero and its variance is unity.

$$E[u] = 0, \qquad E[u^2] = 1 \qquad\qquad (4.5.1)$$

This situation is illustrated schematically in Figure 4.5.1A. The ensemble of membrane potentials is sampled at time t. The average value of the sampled potentials (when the number of traces becomes infinitely large) is zero, and the sampled values are distributed according to a probability density function (p.d.f.), as shown in Figure 4.5.1B. The probability that the membrane potential is above the threshold (T) at time t is given by the area under that part of the p.d.f. that is to the right of T. If the cells could fire, their firing rate would be proportional to that probability [equation (4.2.15)].

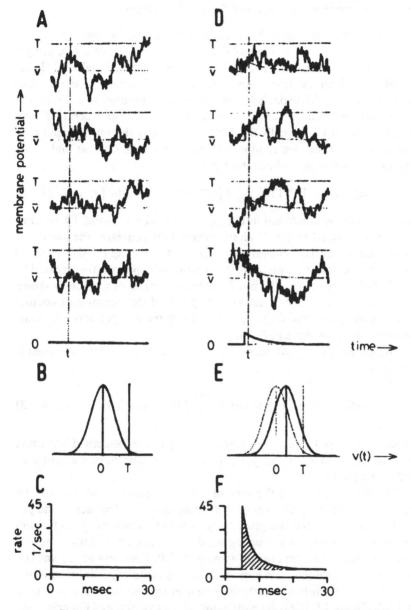

Figure 4.5.1. Membrane potential fluctuations and the probability of firing. A: Four samples of membrane potentials fluctuating around a constant mean; v is the mean, and T is the threshold. B: Probability density function of the membrane potential at time t. C: Expected firing rate of the cell as a function of time. D-F: Same as A-C, but with an EPSP added to the randomly changing membrane potential. [From M. Abeles, "Role of Cortical Neuron: Integrator or Coincidence Detector?" *Israel Journal of Medical Sciences*, 18:83–92, 1982, with permission.]

Imagine that we repeat the same experiment, except that at time zero we generate a small EPSP on all the cells. If at a given time t we sample all the membrane potentials, we have a new random variable u'. If the EPSP is small and we can assume that it is added linearly to all the other processes, then the expected value of the new random variable (u') at time t will be that of the EPSP at that time (bottom trace of Figure 4.5.1D). If for all the cells the EPSP has exactly the same shape, then the variance of the new random variable will be the same as that of the membrane potential without the EPSP.

$$E[u'(t)] = \text{EPSP}(t), \qquad E[u'^2] - E^2[u'] = E[u^2] = 1 \qquad (4.5.2)$$

This situation is illustrated in Figure 4.5.1D, where a small EPSP (dotted line) is added to the random process that generates the simulated membrane potentials. Now the average of all the samples taken at time t is not at zero, but is elevated by the amount of depolarization created by the EPSP at that time. The p.d.f. of this new random variable is shown in Figure 4.5.1E. It looks just like the p.d.f. of the membrane potential without the EPSP, but shifted to the right by the amount of the depolarization caused by the EPSP at t.

The probability of the membrane potential being above threshold is given by

$$\text{pr}\{u' > \theta\} = \int_\theta^\infty f_{u'}(v)\,dv = \int_\theta^\infty f_u[v - \text{EPSP}(t)]\,dv \qquad (4.5.3)$$

where f_u is the p.d.f. of the membrane potentials, as described in the first experiment, and $f_{u'}$ is the p.d.f. of the membrane potential when an EPSP is added to it.

The time t for which the membrane potentials are sampled can be varied, and all the processes of evaluating the expected membrane potential, its p.d.f., and the probability that the membrane potential will be above the threshold can be repeated. This can be done for all possible t values. In the first experiment (where an EPSP was not added and the process was stationary), we should find the same expected membrane potential for all the times (t), as shown in the bottom trace of Figure 4.5.1A. The p.d.f. (f_u) will be the same for all t values, and the firing rate of the cell will therefore be constant at all t values, as shown in Figure 4.5.1C.

If we repeat the same measurements in the second experiment (where an EPSP is added), the expected membrane potential will vary over time. Its shape will follow exactly that of the EPSP that was added, as shown in the bottom trace of Figure 4.5.1D. The p.d.f. ($f_{u'(t)}$) is now moving over time. At first it jumps to the right, and then creeps back exponentially to its resting position. The probability of the membrane potential being above threshold (pr$\{u'(t) > T\}$) also changes over time, and so does the firing rate of these cells.

For every time instant t, we compute the probability that the membrane potential $u'(t)$ is above threshold, and from that probability we evaluate [by equation (4.2.15)] the instantaneous firing rate $\lambda(t)$ at that time. This is done in Figure 4.5.1F, which shows how the firing rate will change if the p.d.f. of the membrane potential has a Gaussian form. The figure shows that the firing rate increases suddenly when the cell becomes depolarized, and then decays back to its resting level. However, the decay is faster than a simple exponential decay. This faster decay can be expected as long as the threshold (T) is situated far out on the decaying limb of the p.d.f. There, the area under the curve (giving the probability that u' is above threshold) falls off faster than in direct proportion to the amount of shift in position caused by the depolarization. This property is true for any p.d.f. that decays gradually to zero for large membrane potentials. The firing rate curve in Figure 4.5.1F shows the expected shape of the cross-correlation curve between a presynaptic neuron and a postsynaptic neuron.

The foregoing argument indicates that for a postsynaptic neuron that does not fire periodically and exhibits a low spontaneous firing rate (T is at the tail of the p.d.f.), and when the EPSP is small, the cross-correlation function will fall off faster than the membrane potential.

For the same reasons, cortical neurons must be very sensitive to synchronous activation of several EPSPs. This sensitivity is based on two premises: that at every time instant, the firing rate is proportional to the probability of the membrane potential being above threshold, and that the p.d.f. declines toward zero in a convex manner (when looked at from below). When these conditions apply, the increment in area under the p.d.f. caused by two superimposed EPSPs is greater than twice the increment caused by one of the EPSPs alone. See Abeles [1982b] for an elaboration of this argument.

The opposite argument holds true for an IPSP. There, the probability of the membrane potential being above threshold is lowered while the cell is hyperpolarized (or while the variance of the membrane fluctuations is reduced by increased permeability to chloride ions). However, the cortical neurons usually fire at low rates, and it is therefore more difficult (statistically speaking) to detect a missing spike than an added spike [Aertsen & Gerstein, 1985].

The experiments described here, in which many neurons have to be measured simultaneously, cannot be carried out in reality. Instead, we must resort to approximations. In the first hypothetical experiment (Figure 4.5.1A), the sampling time bears no relation to events that might affect the membrane potential of the cells. Under this condition, the statistics for every cell are stationary, in the wide sense; that is, the expected membrane potential $E[u(t)]$ and its variance $E[u^2(t)]$ are not functions of time. We assume, further, that the process is ergodic. That is, the statistics obtained by averaging over time will converge to the same expected values that are obtained by averaging across many wave shapes at one time instant. Therefore, by penetrating one cell with a microelectrode, we can estimate its average membrane potential and variance by sampling at many points across time (as described in Section 4.2) and by assuming that these values are the same as those that would have been obtained in the first hypothetical experiment described in this section.

A word of caution is due: Estimates of average membrane potential and variance are always based on a finite sampling time. An evaluation of how true to life these estimates are must be based on what is called in statistics the number of degrees of freedom (which is essentially the number of independent measurements on which our estimates are based). These will not be the number of samples, minus 1 (for the average), or minus 2 (for the variance), because our samples are not independent. To be on the safe side, we must compute the autocorrelation function [equation (4.2.7)] and determine the delay until the correlation drops to zero. Those samples that are as far apart as this delay are uncorrelated, and therefore can be assumed to be independent.

In our second hypothetical experiment (Figure 4.5.1D), we add an EPSP to the random process. The new process is no longer stationary, and therefore we cannot compute the statistical parameters simply by averaging across time. What we do, then, is assume that the perturbation introduced by the added EPSP fades away after a while (say 1 s), so that we can introduce another one, and its effect will be independent of whatever the previous added EPSP caused. We then cut the recorded activity into many short pieces, each of which starts with the EPSP (or at some fixed time before it), and treat them as if they were the ensemble of records in Figure 4.5.1D.

Note that all the parameters dealt with in this chapter (the average membrane potential, its variance, the p.d.f. of the membrane potential, the autocorrelation of the membrane potential, the threshold, the average shape of an EPSP, and the firing rate of the cell) can be measured directly. Therefore, all the relations suggested here can be tested experimentally.

4.5.2 The effect of large synaptic potentials

In the preceding section we dealt with the cell's reaction to a small EPSP; here we analyze how the cell responds to a large EPSP. When we had only a small EPSP, the probability of getting two or more spikes during a given EPSP was very small, and we therefore ignored these cases. However, that is not so for a very large EPSP. Here, the membrane potential usually will reach above threshold and, if the action potential is blocked, will stay there for some time before repolarization lowers it again below threshold. Thus, for a large EPSP, we must consider the effect of one spike on the probability of subsequent firing.

In order to consider this effect, we represent the threshold (θ) as the sum of two values: a constant value (θ_0), representing the threshold at rest, and a threshold change ($\Delta\theta$), representing the effect of previous firing on the threshold.

$$\theta = \theta_0 + \Delta\theta \qquad (4.5.4)$$

We adhere to the idea that the firing rate at a given moment t is proportional to the probability that the membrane potential is above threshold at that moment:

$$\lambda(t) = pr\{v(t) > \theta(t)\}/r \qquad (4.5.5)$$

This is similar to equation (4.2.15), except that now θ varies over time.

We further assume that the effects of the action potentials on the threshold are cumulative, so that every spike (at time t_i) sets up a threshold change function $\Delta(t - t_i)$, and that all these effects are cumulative.

$$\theta(t) - \theta_0 = \sum_{i=1}^{\infty} \Delta(t - t_i) \qquad (4.5.6)$$

This is certainly not the physiological mechanism by which several successive action potentials affect the threshold. However, it is possible to obtain an appropriate Δ function that will approximate the physiological processes phenomenologically.

The probability of finding a spike at time $(t, t + dt)$ is given by $\lambda(t)\, dt$. We can use this to compute the expected threshold at time t:

$$E[\theta(t)] = \int_{-\infty}^{\infty} \lambda(\tau)\Delta(t - \tau)\, d\tau + \theta_0 =$$

$$\int_{-\infty}^{\infty} \lambda(t - \tau)\Delta(\tau)\, d\tau + \theta_0 = \int_{0}^{\infty} \lambda(t - \tau)\Delta(\tau)\, d\tau + \theta_0 \quad (4.5.7)$$

Changing the integration limits in the last term is allowed because the spike cannot affect the threshold before it occurs, and therefore $\Delta(\tau)$ has zero value for negative times.

If we know the shape of the EPSP and the statistics of membrane fluctuations around this shape, we can compute the probability of the membrane potential being above the threshold (and from that the firing rate) by

$$\lambda(t) = \int_{\theta}^{\infty} f_{V(t)}(v)\, dv/r \quad (4.5.8)$$

where $f_{V(t)}$ is the p.d.f. of the membrane potential at time t, and θ is the *expected* threshold at time t [i.e., $\theta = E[\theta(t)]$, given by equation (4.5.7)].

We obtain the two equations (4.5.7) and (4.5.8) with two variables (θ and λ) that can be solved. The strange appearance of $\lambda(t)$ in a convolution integral in equation (4.5.7) is not a real hindrance. The current threshold (at t) is a function of the firing rate in the past only [equation (4.5.7)], whereas the firing rate at the next time instant will depend only on the current threshold [equation (4.5.8)]. Therefore, by starting off with a steady-state condition (where $\theta = \theta_0$), and chopping time into very small steps, we can successively apply equations (4.5.8) and (4.5.7) to solve the time evolution of the firing rate and threshold.

Problem 4.5.1. Compute the firing rate of a cell when an EPSP is introduced at time zero. Assume the following conditions: (1) The membrane potential behaves like a random Gaussian variable. (2) The resting firing rate is five spikes per second. (3) The EPSP looks like

$$\text{EPSP}(t) = 10^3 t e^{-t/0.005}$$

(4) It is summed linearly with the membrane potential. (5) The threshold, after firing, looks like

$$\theta(t) = 10 + (10^{-6}/t^3)e^{-t/0.001}$$

(where *t* is given in seconds and the membrane voltage is in millivolts).

After computing the cell's response to this EPSP, gradually reduce the amplitude of the EPSP. Does the response approach the shape predicted in Section 4.5.1?

On a qualitative level, the argument of Knox (Section 4.4) can also be applied to the effect of a large EPSP on a randomly firing neuron. If the membrane potential when the EPSP starts is depolarized enough, the cell is likely to fire during the rising phase of the EPSP, and then become refractory for a few milliseconds and not fire again. If the EPSP happens to be superimposed on a deeply hyperpolarized membrane potential, so that even at its peak it does not reach the threshold, it is not likely that the membrane potential will reach threshold during the falling phase of the EPSP. Thus, most of the postsynaptic spikes generated by a large EPSP will be concentrated during its rising phase.

A large IPSP, on the other hand, will depress the firing rate of the randomly firing neuron for most of its duration.

To summarize, cortical neurons, in most areas, act like the randomly firing neuron whose firing rate is proportional to the probability of the membrane potential being above threshold. Most EPSPs in the cortex are small and generate an elevation in the cross-correlation that looks somewhat like an EPSP, but with a shortened falling phase and with an additional delay of about 1 ms. IPSPs depress the firing rate for their entire duration, but are very difficult to detect by cross-correlations. Large EPSPs tend to have an effect that is concentrated around the rising phase of the EPSP.

4.6 References

Abeles M. (1982a). *Studies of Brain Function. Vol. 6: Local Cortical Circuits: An Electrophysiological Study.* Springer-Verlag, Berlin.
(1982b). Role of cortical neuron: Integrator or coincidence detector? *Isr. J. Med. Sci.* 18:83–92.
Aertsen A. M. H. J. and Gerstein G. L. (1985). Evaluation of neuronal connectivity: Sensitivity of cross correlation. *Brain Res.* 340:341–54.
Ahissar E. and Vaadia E. (in press). Oscillatory activity of single units in a somatosensory cortex of an awake monkey and their possible role in texture analysis. *Proc. Natl. Acad. Sci. U.S.A.*
Calvin W. H. (1974). Three modes of repetitive firing and the role of threshold time course between spikes. *Brain Res.* 69:341–6.
Calvin W. H. and Stevens C. F. (1968). Synaptic noise and other sources of

randomness in motoneuron interspike intervals. *J. neurophysiol.* 31:574–87.

Cope D. K. and Tuckwell H. C. (1979). Firing rates of neurons with random excitation and inhibition. *J. Theor. Biol.* 80:1–14.

Cox D. R. (1962). *Renewal Theory.* Methuen, London.

Cox D. R. and Isham V. (1980). *Point Processes.* Chapman & Hall, London.

Eccles J. C. (1957). *The Physiology of the Nerve Cell.* Johns Hopkins University Press, Baltimore.

Eckhorn R., Bauer R., Jordan W., Brosch M., Munk M., and Reitboeck H. J. (1988). Coherent oscillations: A mechanism of feature linking in the visual cortex? *Biol. Cybernet.* 60:121–30.

Ellul R. (1972). The genesis of the EEG. *Int. Rev. Neurobiol.* 15:227–72.

Fetz E. E. and Gustafsson B. (1983). Relation between shapes of post-synaptic potentials and changes in firing probability of cat motoneuron. *J. Physiol. (Lond.)* 341:387–410.

Glenn L. L. and Dement W. C. (1981). Membrane potential, synaptic activity, and excitability of hindlimb motoneurons during wakefulness and sleep. *J. Neurophysiol.* 46:839–54.

Gray C. M. and Singer W. (1989). Stimulus-specific oscillations in orientation columns of cat visual cortex. *Proc. Natl. Acad. Sci. U.S.A.* 86:1698–702.

Holden A. V. (1976). *Lecture Notes in Biomathematics. Vol. 12: Models of the Stochastic Activity of Neurons.* Springer-Verlag, Berlin.

Kirkwood P. A. (1979). On the use and interpretation of cross-correlation measurements in the mammalian central nervous system. *J. Neurosci. Meth.* 1:107–32.

Kirkwood P. A. and Sears T. A. (1982). The effects of single afferent impulses on the probability of firing of external intercostal motoneurons in the cat. *J. Physiol. (Lond.)* 322:315–36.

Knox C. K. (1974). Cross-correlation function for a neuronal model. *Biophys. J.* 14:567–82.

(1981). Detection of neuronal interactions. *TINS* 4:222–4.

Knox C. K. and Popple R. E. (1977). Correlation analysis of stimulus-evoked changes in excitability of spontaneously firing neurons. *J. Neurophysiol.* 40:616–25.

Komatsu Y., Nakajima S., Toyama K., and Fetz E. E. (1988). Intracortical connectivity revealed by spike-triggered averaging in slice preparations of cat visual cortex. *Brain Res.* 442:359–62.

Lansky P. (1984). An approximation of Stein's neuronal model. *J. Theor. Biol.* 107:631–47.

Perkel D. H., Gerstein G. L., and Moore G. P. (1967). Neuronal spike trains and stochastic point processes. I. The single spike train. *Biophys. J.* 7:391–418.

Rall W. (1962). Electrophysiology of a dendritic neuron model. *Biophys. J.* 2:145–67.

Rice S. O. (1954). Mathematical analysis of random noise. In Wax N. (ed.), *Selected Papers on Noise and Stochastic Processes*, pp. 133–294. Dover, New York; reprinted from *Bell Sys. Tech. J.*, Vol. 23–4.

Sampath G. and Srinivasan S. K. (1977). *Lecture Notes in Biomathematics. Vol. 16: Stochastic Models for Spike Trains of Single Neurons.* Springer-Verlag, Berlin.

Stein R. B. (1965). A theoretical analysis of neuronal variability. *Biophys. J.* 5:173–94.

Tuckwell H. C. and Cope D. K. (1980). Accuracy of neuronal interspike times calculated from a diffusion approximation. *J. Theor. Biol.* 83:377–87.

Woody C. D. and Black-Cleworth P. (1973). Difference in excitability of cortical neurons as a function of motor projection in conditioned cats. *J. Neurophysiol.* 36:1104–16.

5
Models of neural networks

5.1 Introduction

The methods and data presented in Chapters 2, 3, and 4 are essential for assessing the feasibility of neuronal circuits composed of a small number of elements. However, the neural network models presented in recent years have demonstrated that computations can also be carried out by massive interactions among a multitude of neurons. This chapter offers an introduction to current trends in neural network modeling.

Modeling of neural networks has been carried out extensively in the past five years. Among the first attempt to build circuits that would compute were McCullouch and Pitts [1943], who showed how to compute logic predicates with neurons. They later constructed a neuronal circuit that recognized shapes, regardless of their position in the visual field [Pitts and McCulloch, 1947]. Subsequently there were many other attempts to construct computing circuits from neural-like elements [e.g., Wooldridge, 1979], but most of those attempts did not leave a lasting impression in the neurosciences. The original McCulloch and Pitts paper was difficult to follow, even for mathematicians [Palm, 1986b]; thus, adoption and development of their ideas by neurophysiologists did not follow. The image-recognizing circuit they developed had connectivities that, so far as we know, did not exist in the visual areas. That seems to have been the fate of most "neural computers" suggested in the past.

Following the publication of Sholl's book [1956] on the quantitative anatomy of the visual cortex there was an outburst of theoretical works on random networks [Beurle, 1956; Caianiello, 1961; Griffith, 1963, 1971; Anninos et al., 1970; Harth et al., 1970; Wilson and Cowan, 1973]. Although the community of experimental neurophysiologists lost interest in the subject after a while, some interesting properties were embodied in those early models. One such feature relates to the question of stable levels of "spontaneous" activity exhibited by cortical neurons.

150

The properties of such random networks are discussed in Sections 5.3 and 5.4, where it is shown that one experimental finding – that even without any apparent external sensory stimulus, cortical neurons continue to fire at low rates (in an apparently random fashion) – is by no means trivial.

In recent years there has been a revival of interest in the theory and modeling of neural networks that compute. Two of the prevailing models – the recurrent neural networks and the multilayered perceptrons – will be introduced in the last sections.

Notation. The following notation is used throughout this chapter:

σ_i	The spiking of neuron i [on a $(0, 1)$ scale]; $\sigma_i = 1$ if neuron i fired, and σ_i if neuron i did not fire.
σ	The state of the network given by the list of spiking states of all the neurons [e.g., $\sigma = (1, 1, 1, 0, 0, \ldots, 0)$ means that neurons 1, 2, and 3 fired, but all the others were quiescent].
S_i	The spiking of neuron i [on a $(-1, 1)$ scale]. $S_i = 1$ if the neuron fired, and $S_i = -1$ if not.
S	The state of the network given by the list of spiking states S_i of all the neurons [e.g., $S = (1, 1, 1, -1, -1, \ldots, -1)$ means that neurons 1, 2, and 3 fired, but all the others were quiescent].
ξ, ξ^μ	A vector of ones and zeros (or -1's) describing a desired firing state of the entire network (i.e., a particularly desired σ). Usually such a desired state represents a memory that the network ought to recall under some appropriate conditions.
P	The number of memories (ξ^μ) embedded in the network.
$\theta, \theta_i, \theta_i(t)$	Threshold, threshold of neuron i, and threshold of neuron i at time t (when spiking is denoted by σ).
$T, T_i, T_i(t)$	Threshold, threshold of neuron i, and threshold on neuron i at time t (when spiking is denoted by S).
$U, U_i, U_i(t)$	The membrane potential at the "trigger zone" of the action potential, the membrane potential of neuron i, and the membrane potential of neuron i at time t.
J_{ij}	The strength of the (synaptic) junction from neuron j to neuron i. J_{ij} is scaled in such a way that the contribution of the presynaptic firing of neuron j to the membrane potential of neuron i is given by $J_{ij} \cdot \sigma_j$. J_{ij} is positive for an excitatory synapse, and negative for an inhibitory synapse.
J	The matrix whose elements are the junction strengths J_{ij}.
$D(S)$	A discontent function of a network (when at state S) showing to what extent the postsynaptic neurons follow the effects of their presynaptic inputs.
N	The number of neurons in the network.

5.2 The point neuron

In most theories and simulations discussed here, a highly simplified model of a neuron is used. The neuron has the following features:

1. When a presynaptic spike (σ_j) arrive at a synapse (J_{ij}), after a delay of one time unit it depolarizes (hyperpolarizes) the membrane potential of the postsynaptic cell (U_i) by a fixed amount given by $\sigma_j \cdot J_{ij}$. That is, after the presynaptic cell number j fires ($\sigma_j = 1$), then the membrane potential of the postsynaptic cell changes by J_{ij}. If the synapse from neuron j to neuron i is excitatory J_{ij} is positive, but if it is inhibitory, J_{ij} is negative.

2 The synaptic effects last for only one time unit.

3. All the synaptic effects add up linearly. Therefore, we can write

$$U_i(t) = \sum_{j=1}^{N} J_{ij} \cdot \sigma_j(t - 1) \tag{5.2.1}$$

Note that when there is no activity ($\sigma_j = 0$ for all j), the membrane potential (U_i) is zero. Thus, in this representation the resting potential is called zero.

4. The neuron fires whenever its membrane potential reaches the threshold.

$$\sigma_j(t) = \begin{cases} 1 & \text{for } U_i(t) > \theta_i(t) \\ 0 & \text{otherwise} \end{cases} \tag{5.2.2}$$

5. The cell is refractory for one time unit. In some of the models it is assumed that the neuron is refractory for two time units. This assumption can be reconciled with equation (5.2.2) if we assume that after firing, $\theta_i(t)$ is very high for an additional time unit.

6. No neuron synapses on itself.

$$J_{ii} = 0 \quad \text{for all } i$$

Different models require minor variations of this basic model. These will be described in the appropriate sections.

The "point neuron" differs in many ways from real neurons. It does not allow for modulation of synaptic strength by processes such as facilitation or post-tetanic potentiation. It does not admit the possibility of nonadditive interactions among synaptic inputs. It does not allow for inhibition by shunting the membrane potential. It does not exhibit gradual recovery of the membrane potential in the direction of the resting value. It does not allow for inclusion of the relative refractory period, nor does it allow for accommodation of the threshold. Some of these restrictions have been relaxed for some of the models, without qualitatively changing the properties of the model. However, there is no assurance that if all these restrictions were relaxed at the same time, the main results of the models would remain valid. A more detailed discussion of point neurons can be found in Chapters 2, 14, and 15 of MacGregor's book [1987] on neural modeling. We should note that there have been few analytical studies pertaining to networks composed of neurons with more natural properties. Also, there have been few simulations of large networks of neurons with natural properties, because most researchers cannot afford the computer resources required.

5.3. Small random networks

In this section we follow the analysis of Harth et al. [1970] and Anninos et al. [1970] concerning "netlets" that consist of point neurons connected to one another by a few synapses (on the order of tens) and having low thresholds (in terms of how many EPSPs are required to fire the neuron). Although the anatomy of the cortex suggests that the connectivity in the cortex is much richer than is assumed here, one may claim that there are only a few strong connections, so that each piece of cortex may be considered to be composed of many netlets, with strong coupling among the members of each netlet, and weaker coupling between netlets [Harth et al., 1970]. Studying the behavior of such netlets also provides insight for understanding large random networks.

The analysis of Anninos and Harth assumes that time is described in units that are equal to one synaptic delay (about 1 ms), that the refractory period is between one and two time units, and that the duration of the synaptic effect is shorter than one time unit. At time zero, few cells are fired (by an "external" source), and then the state (firing/not firing) of each of the neurons is examined at times 1, 2, and so forth. This form of simulating a network is called synchronous-parallel updating. Parallel updating means that the states of all neurons are checked (and updated if necessary) before they have a chance to exert their influence (i.e., within one synaptic delay). Synchronous updating means that the simulation is carried out as if all the discharges occur synchronously at times 1, 2, 3, . . . ms, but not in between.

This selection of parameters and operating mode has the advantage that the activity of the "net" at a given time step depends only on its

A   B

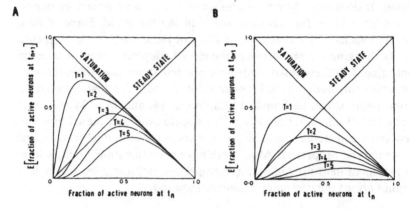

Figure 5.3.1. Behavior of small netlets. The graphs show the relationship between the number of neurons that fired at time step n and the expected number of firing neurons at time step $n + 1$. The straight line $y = x$ represents steady states, in which equal numbers of neurons fire at both time n and time $n + 1$. The straight line $y = 1 + x$ represents saturation, in which every neuron that is not refractory (because it fired at time n) fires at time $n + 1$. A: All neurons are excitatory. A stable ongoing activity can be reached only near saturation. B: 30 percent of the neurons are inhibitory. When the lowest level of *stable* ongoing activity is attained, every neuron has a firing probability of 0.25. If one step corresponds to 1 ms, this means that the firing rate is 250 spikes per second. [Adapted from P. A. Anninos, B. Beek, T. J. Csermely, E. M. Harth, and G. Pertile, "Dynamics of Neural Structures," *Journal of Theoretical Biology,* 26:121–28, 1970, with permission of Academic Press Inc.]

activity in the previous time step. By averaging over all the possible netlets that obey the same statistical rules of connectivity, we can obtain graphs (Figure 5.3.1) that describe the expected number of firing neurons at step $n + 1$ as a function of the number of neurons that fired at the nth step.

Figure 5.3.1A shows the expected behavior of a netlet containing only excitatory neurons. Although the number of neurons in the netlet is assumed to be very large, each neuron is connected to only ten other neurons in the netlet. The synaptic strength is 1. In the present model, cells that fire at time step n cannot fire again at time step $n + 1$. Therefore, the sum of the firing cells at steps n and $n + 1$ cannot be greater than the total number of neurons in the netlet. This means that the netlet's state cannot extend beyond the line $x + y = 1$ in Figure 5.3.1. If the netlet reaches a steady state, the number of active neurons at step n

+ 1 should be equal to the number of active neurons at the nth step. Thus, steady states are described on the line $y = x$.

The set of graphs shown in Figure 5.3.1A shows the expected behavior of the netlet when the threshold is set at the amplitude of 1, 2, 3, 4, or 5 EPSPs. The graph corresponding to $T = 1$ is unique in that it is the only one that starts with a positive slope (all the others start with a slope of zero). Obviously there is a stable point at (0, 0). If no neuron fires at the nth step, then the netlet remains absolutely quiescent at step $n + 1$. If even one neuron fires, then at the second step ten neurons will fire (all those that receive an excitatory synapse from the previously firing neuron), at the third step almost 100 neurons will fire (all those that are connected to the ten neurons that fired at step 2 and are not refractory), and so on until the netlet reaches saturation (i.e., until at every step all the neurons are either firing or refractory). That happens at the point where the graph crosses the steady-state line, at (0.5, 0.5). In this state, every neuron fires on every other time step (i.e., at about 500/s).

All the other graphs start with zero slope, because when the activity at the nth time step is very low (e.g., only one neuron fires), no neuron reaches threshold, and therefore the activity at step $n + 1$ is zero. When the threshold is high ($T = 4$ or $T = 5$), the entire graph lies under the steady-state line. In these netlets, regardless of how many neurons are activated at time zero, the activity will soon fade and reach a stable quiet state at (0, 0).

At the intermediate states ($T = 2$ and $T = 3$), the graphs meet the steady-state line at three points. At (0, 0) we have a stable steady state in which all neurons are quiescent. The graph starts with a zero slope, but then climbs up to cross the steady-state line from below. This is an unstable steady state (to be described shortly), and then the graph bends down, crosses the steady-state line again, and approaches the saturation line asymptotically.

Figure 5.3.1B shows the behavior of a similar netlet in which 30 percent of the neurons are inhibitory. Again, each neuron has synapses to ten other neurons, the strength of the excitatory synapse is 1, and that of the inhibitory synapse is -1. The overall behavior seen in Figure 5.3.1B is similar to that seen in Figure 5.3.1A, except that the corresponding curves are at lower levels. Thus, the addition of inhibition does not alter the behavior of the netlet qualitatively. Note that even at the lowest curve ($T = 2$) that has a stable activity, the firing rate of the neurons during the steady state is approximately 250/s.

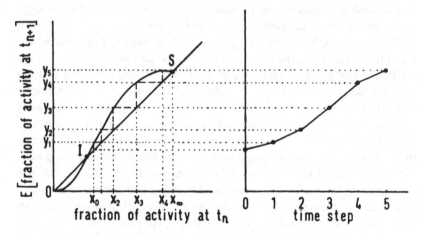

Figure 5.3.2. Evolution of activity with time. On the left there is a graph similar to that of Figure 5.3.1. Of the three putative steady states of 0, I, and S, only 0 and S are stable. If the activity is slightly above that of point I, it will increase until reaching point S. On the right, the evolution of activity with time is shown. [Adapted from P. A. Anninos, B. Beek, T. J. Csermely, E. M. Harth, and G. Pertile, "Dynamics of Neural Structures," *Journal of Theoretical Biology*, 26:121–48, 1970, with permission of Academic Press Inc.]

The expected evolution of activity over time can be studied in a way similar to that shown in Figure 5.3.2. On the left we see a graph of the expected fraction of firing neurons at step $n + 1$ as a function of the fraction of neurons that fired at the nth step. The three putative steady states are marked 0, I, and S. Let us examine what happens when we start with activity level X_0, which is slightly above the activity level of I. The behavior of the network is characterized by the point (X_0, Y_1), at which the expected level of activity Y_1 is slightly above the eliciting activity X_0. If in the first step the net fires as expected (i.e., if $X_1 = Y_1$), then in the second step the behavior is represented by (X_1, Y_2), in which Y_2 is still higher. If the network again behaves as expected $(X_2 = Y_2)$, then the next expected activity Y_3 goes still higher, and so on until the activity settles at the X_∞ level that corresponds to point S. We can see these dynamics by plotting the activity levels versus the time steps, as is done on the right in Figure 5.3.2. The reader can verify that if we start at activity levels (X_0) that are below point I, the activity will gradually subside until the netlet reaches the stable quiescent point at 0. Thus, point I represents an unstable ignition point for the netlet. If the activity is aroused, by external excitation, above this ignition point, it will increase by itself up to the stable steady-state point S. The netlet will then remain there until driven back below I by some external effect.

The effects of external drive on the netlet can be described in the following manner [Anninos et al., 1970]: Let us add some external excitatory input to all the neurons in the netlet. The behavior of the netlet, which was described by the $e = 0$ graph in Figure 5.3.3A, is now shifted to the graph e_1. The three crossing points 0, I, and S are now shifted to 0_1, I_1, and S_1. Two interesting changes should be noted, First, the 0_1 stable state is no longer at zero activity. Thus, the netlet can maintain a low level of stable ongoing activity. This level is derived principally from the external excitatory input that may activate some of the neurons even when the netlet has shown no previous activity. Second, as the ignition point I_1 is now lower than I, it is easier to drive the network into its active stable steady state S_1.

The changes in the netlet's behavior generated by the external input can be described by a "phase diagram" showing the positions of the crossing points O, I, and S as a function of the level of external input, as shown in Figure 5.3.3B. There, the levels at which the steady-state points 0 are achieved as a function of the external input are plotted as the lower (solid) arm of the S-shaped phase transition boundary. The levels of activity at which the unstable ignition points I are reached as a function of the external input are plotted as a broken line, and the levels for the S stable points are plotted as the top (solid) arm of the phase transition boundary. If we hold the level of activity constant at e_1 and start with a netlet activity that is below I_1, then the activity will subside to the lower steady state 0_1. If, on the other hand, we start from levels that are above I_1, the activity will increase until the high steady stable state S_1 is reached. The small arrows in Figure 5.3.3B indicate in which direction the activity of the netlet will change for all possible initial conditions.

Let us now examine the behavior of the netlet given a changing external input (Figure 5.3.3C). We begin with no external input, when the netlet is at rest at point 0, and slowly increase the external input. The resting state will exhibit a stable low level of activity that will increase only slightly with the increasing external input. This will continue until the external input reaches the level of e^+, at which the low-level steady state will merge with the ignition point, and the netlet will move quickly to the almost saturated steady state in which most of the neurons either fire or are refractory. Then, even if the external input drops back to zero, the netlet will remain at its highly active state S.

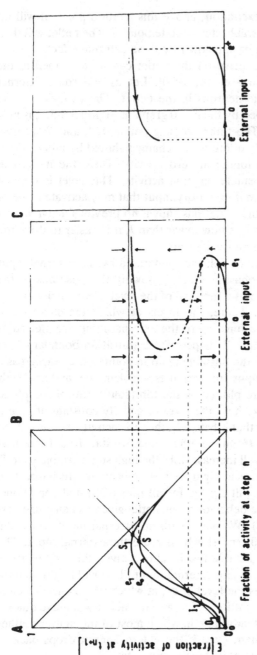

Figure 5.3.3. Effects of external drive on a netlet. A: The netlet behavior with no external drive is marked e_0. When an excitatory drive is added, the behavior shifts to the e_1 curve. The three putative steady states, 0, I, and S are shifted to 0_1, I_1, and S_1. B: The loci of points 0, I, and S as functions of the external drive. The loci of the stable steady states (0 and S) are described by the solid lines, and the locus of the unstable state (I) is described by the broken line. If the network starts from a point that is not on the steady-state line, while keeping the external drive fixed, the network's activity will move toward a steady state. The arrows describe the direction of this movement. C: Description of the behavior of the network when starting off with no external drive, then gradually increasing the input to e^+, then decreasing it to e^-, and finally returning to zero external drive. [Adapted from P. A. Anninos, B. Beek, T. J. Csermely, E. M. Harth, and G. Pertile, "Dynamics of Neural Structures," *Journal of Theoretical Biology*, 26:121–48, 1970, with permission of Academic Press Inc.]

We see that a transient increase in external input throws the netlet from the stable quiescent state into the stable highly active state. If we wish to turn the netlet off, we must add inhibitory external input. At low levels of inhibitory external input, the activity of the netlet is reduced only slightly, but as the inhibition reaches the level of e^-, the active steady state merges with the ignition point, and the netlet is shut off completely. At this point, returning the external input back to zero will leave the netlet at its quiescent steady state.

Although the properties of neurons and synaptic interactions assumed in these models are very restrictive, it seems that the basic properties of the netlets remain valid even when these restrictions are relaxed. However, when more realistic properties are assumed, often one must resort to simulations to analyze the behavior of the model.

A word of caution is due: The behavior described in this section was derived by computing statistical expectations. In other words, this is what would be obtained by averaging the results over all possible netlets that obey the connectivity and excitability rules assumed by the model. Any given netlet may exhibit a very different behavior. For example, it may show cyclic activity or stable activity in which some of the neurons continuously fire at a high rate, whereas others are quiescent. When studying the expected behavior of a class of random netlets (with some statistical rule of conductivity) by simulations, it is a common practice to randomly change all the connections among the neurons after each simulation step [Takizawa and Oonuki, 1980]. Although this procedure yields the desired statistical estimation, it by no means describes how a given netlet with a fixed set of connections will behave.

To summarize, a netlet with excitatory and inhibitory neurons having similar thresholds and connections exhibits two stable operating points: one at zero activity, and one at very high activity. In between these points there is an ignition point. Activity above the ignition point will drive the netlet to its high stable point, whereas activity below the ignition point will relax the netlet to the quiescent state. External input can maintain the netlet at a stable low level of activity. By modulating the external activity, one can switch the netlet on and off.

Problem 5.3.1. Is it possible to design a netlet with no external input that has a stable firing rate at levels comparable to what is seen in the cortex (i.e., five to ten spikes per second)?

5.4 Large random networks

A large random network is a network composed of many (tens of thousands) point neurons that are extensively interconnected by some probabilistic laws. In the simplest random network there is no differentiation according to the type or position of the neurons, so that all pairs of neurons have the same probability of being connected. Although random networks cannot compute, and therefore represent a very poor approximation of the cortex, studying their properties can shed interesting light on cortical mechanisms. The main purpose of this section is to expose the problem of how the cortex maintains a stable low firing rate for most of its neurons most of the time.

We begin our discussion of the properties of large random networks with a brief description of work dealing with the possibility of constructing a network that will excite itself in such a manner that it will maintain a low level of ongoing activity. We remind the reader that in the cortex, single neurons fire "spontaneously," in the absence of an apparent external stimulus, at low rates, as shown in Figure 4.3.1.

We saw in Section 5.3 that without external input, a netlet can maintain only fairly high levels of stable ongoing activity. We shall see that it is not obvious that larger networks can maintain low levels of activity.

5.4.1 A review

Beurle [1956] probably was the first to analyze quantitatively the behavior of a large random network of excitatory neurons. The neurons of Beurle were more realistic than the point neurons described in Section 5.2, as they had long EPSPs that could assume any shape over time and had long refractory periods. The structure of the net itself was also more realistic than in most models, as the probability of contact between two neurons was not homogeneous, but rather fell off exponentially with the distance between the neurons. The firing states of the neurons were updated in the parallel-synchronous mode. Beurle was interested principally in mechanisms that generated waves propagating through the network, and with learning rules that could control the direction of propagations.

While investigating the basic properties of this network, Beurle showed that one cannot obtain a uniform, steady low level of activity in such a network. A network of excitatory neurons can be maintained stable either in the quiescent state (when no neuron fires) or in the

saturated state (when almost every neuron is either firing or refractory).
Beurle did not include inhibitory neurons in his network and had to
resort to an "external controller" that rendered part of the network
refractory whenever the level of ongoing activity began to rise.

The question of the stability of "spontaneous" activity was taken up
again by Griffith [1963], who used point neurons with synchronous-
parallel updating. He again showed that an exclusively excitatory network could not maintain stable partial activity. By adding inhibition,
Griffith obtained stability when each neuron had a firing probability of
0.5. That is, the average firing rate of a neuron in the net was 500/s.
Griffith stated that by adjusting the balance between inhibition and
excitation, one could obtain stable activity at lower levels. Examination
of Figure 5.3.1B shows that this can indeed be achieved, but at firing
rates of over 200/s.

Gaudiano and MacGregor [personal communication] set up a network
of excitatory and inhibitory neurons. By carefully trimming the connectivities between the populations, they achieved stable ongoing activity
at rates of about 150/s in the excitatory neurons and less than 100/s in the
inhibitory neurons. However, they stated that stable low levels of activity
could also be obtained. Pantiliat [1985] attempted to set up a network of
1,000 excitatory and inhibitory neurons. She was unable to find conditions under which the neurons in the network would fire stably at low
rates (about 16/s). She could stabilize the network by introducing nonphysiological inhibitory feedback, which acted with zero delay.

It seems to be very difficult to stabilize a network of inhibitory and
excitatory neurons at very low levels of activity. The source of the difficulty is in the extrasynaptic delay involved in activating the inhibitory
neurons. Indeed, Wilson and Cowan [1973] described the behavior of a
neural network by sets of differential equations. They assumed no
synaptic delay, and then found the range of parameters for which the
network exhibited stable levels of ongoing activity. But even their model
exhibited metastable behavior in which small disturbances set up waves
of activity in space and in time.

5.4.2 Large networks: A look from within

In the following discussion we adopt the approach of Nelken
[1985, 1988], who took a fresh look at the stability of the statistics of
firing of a large network by describing what occurs within one neuron in
such a network.

Let us examine the behavior of a neuron that is embedded in a randomly connected network. Each neuron is assumed to have many inputs, and the activities of the inputs are either uncorrelated or weakly correlated. The neuron itself is assumed to reset its internal state every time it fires. Thus, the statistics of the firing times of such a neuron, when subject to a random independent Poissonian synaptic bombardment, will follow those for a renewal process (i.e., the probability of firing at any time will depend only on the time elapsed since the last firing time, not on what occurred earlier). See Cox [1962] for a definition and discussion of the properties of renewal processes.

Unlike the case of the point neuron (Section 5.2), the synaptic inputs considered in Nelken's neuron are quite realistic: The effects can last for variable times, they may interact nonadditively, shunting inhibtions are allowed, and so forth. The only two restrictions on synaptic properties are that the synaptic inputs can be divided into subgroups, so that within each subgroup the effects are additive, and the effect of each type of synapse does not vary over time (e.g., a synaptic effect cannot be inhibitory for the first few milliseconds and then become excitatory).

Nelken was able to show that if the groups of additive synapses contain fifty or more presynaptic sources, and if all of the firing times of every one of the presynaptic neurons follow the statistics of a renewal process, then the synaptic bombardment of the entire group will be practically indistinguishable from a Poisson process. This remains true even if the inputs are weakly correlated pairwise.

Thus, from within, the neuron "sees" Poissonian synaptic bombardment at each of the groups of identical synapses. Note that the interactions between the groups may be highly nonlinear. With this type of input, the output of the neuron (its firing times) will also follow the statistics of a renewal process. This holds true for every neuron in the network. *Therefore, the renewal statistics describe a consistent mode of firing in such a network.*

This result shows that for a very large network that is not subject to strong external drive, the detailed statistics of firing times, as seen in the renewal density (autocorrelation) function of the neuron, reflect only the internal excitability changes of the firing neuron and are insensitive to the firing patterns of the presynaptic neurons that drive it. The only parameter that is affected by the presynaptic drive is the average firing rate of the postsynaptic neuron.

The behavior of each of Nelken's neurons can be described by an input–output curve showing the rate of synaptic bombardment on the

A   B

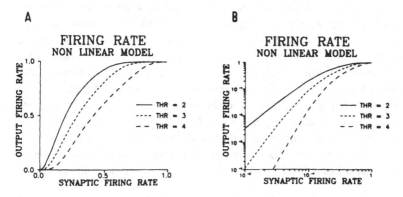

Figure 5.4.1. Firing rate of a postsynaptic (output) neuron as a function of the averaged firing rate of each presynaptic (input) neuron. The graphs display the exact solution for a neuron having nine excitatory inputs and two inhibitory inputs interacting in a nonlinear (multiplicative) manner. THR is the distance between the resting membrane potential and threshold. A: On a linear scale. B: On a double logarithmic scale. At low firing rates, the graphs look like $y = x^T$. [Based on I. Nelken, "Analysis of the Activity of Single Neurons in Stochastic Settings," *Biological Cybernetics*, 59:201–15, 1988, with permission.]

abscissa and the rate of firing of the postsynaptic neuron on the ordinate (Figure 5.4.1). At low firing rates, the relation is convex (bending upward), and if the threshold is T EPSPs above the average membrane potential, the curve looks like x^T. Thus, an isolated EPSP has a negligible probability of evoking a spike. Rather, a spike will be evoked when T EPSPs arrive together. This feature of the curve illustrates the sensitivity of the neuron to coincident activation (a topic that is elaborated extensively in Chapter 7).

Let us now turn back to the question of stable ongoing activity within a network of excitatory (E) and inhibitory (I) neurons. The following discussion gives some insight into why it is so difficult to stabilize such a network at a low level of ongoing activity.

We assume that within the network the excitatory (E) neurons excite both the other excitatory neurons and the inhibitory (I) neurons, whereas the inhibitory neurons inhibit only the excitatory neurons. In other words, the strength of the synapses from an I neuron to an I neuron is set at zero. We allow ourselves to set the probabilities of finding a synapse from an E neuron to another E neuron (from E to I *and from I to E*) and to set the synpatic strength of these connections at any desired value.

We continue to analyze the behavior of the network by observing one E neuron from within. To simplify the analysis, we assume that the postsynaptic potentials are short (lasting only 1 ms), and we update the network in the parallel synchronous mode. We can now construct a figure similar to Figure 5.4.1, but with two differences. In Figure 5.4.1 the steady-state firing rate of the neuron was plotted as a function of the steady-state firing rate of its inputs. Now we plot the expected firing rate of an E neuron at time step $n + 1$ as a function of the average firing rate of the rest of the E neurons at time step n. In Figure 5.4.1, a neuron with a given threshold value had only one input–output curve, whereas now each neuron has a family of curves – each for a different firing rate of the I neurons.

Plots of this type are shown in Figure 5.4.2, through which we can investigate the stability of the network. We see there the firing rate of an excitatory neuron shown as a function of the firing rate of the other excitatory neurons in the network. Each of the curves describes a behavior of the excitatory neuron under a different fixed level of inhibition: $I(0)$ for the network when the inhibitory neurons are quiescent, $I(E_0)$ when the inhibitory neurons are at the level associated with the desired ongoing activity, $I(E_1)$ and $I(E_2)$ for progressively increasing levels of inhibition. Following Nelken's results, all the curves look like x^T, where T is the threshold level expressed as the number of EPSPs required to reach it. In Figure 5.4.2, T is set at eight EPSPs. The graphs are drawn on logarithmic scales and therefore appear as straight lines, with a slope of T ($\log x^T = T \log x$). At low firing rates, all the curves must cross the steady-state ($y = x$) line in the manner shown in Figure 5.4.2.

When the rate of the excitatory neurons is at E_0 (5/s) for a long time, the rate of the inhibitory neurons is $I(E_0)$, and the net can continue indefinitely in this manner as long as no fluctuations around this state occur. Let us assume that at time step 1 the overall activity of the excitatory neurons increases slightly to E_1 (6/s in Figure 5.4.2B). At the next time step the level of activity of the excitatory neurons will go to E_2 (21/s), which is much greater than E_1. The level of activity of the inhibitory neurons will also increase, but the effect will be noticeable only at the next step (there is one synaptic delay in the E-to-I connection and one more delay in the feedback I-to-E synapses). Therefore, for time step 2, we move from the $I(E_0)$ curve to the $I(E_1)$ curve (Figure 5.4.2C). Then the activity of the excitatory neurons will drop slightly to E_3 (15/s). At time step 3 (Figure 5.4.2D), we move to the $I(E_2)$ curve, which is farther to the right because $E_2 > E_1$, and therefore the inhibitory activity at time step 2 must be greater than at time step 1. The excitatory activity will fall to E_4 (0.3/s). Note that had $I(E_3)$ moved farther to the right, E_4 would have fallen to an even lower activity. Thus, at time step 4, we face a very low firing rate, but the inhibition is still very active (because it depends on the excitatory activity E_3, which was quite high). $I(E_3)$ must lie between $I(E_2)$ and $I(E_1)$, so that at step 5 the excitatory activity will fall way down to about 10^{-14}/s, from which it cannot recover.

Thus, E_0 is not a stable steady state for this network. By playing around with graphs such as $I(E_0)$ and $I(E_1)$ in Figure 5.4.2, we can see that it is difficult to obtain a random network of excitatory and inhibitory neurons that has a stable steady state. Note that had we allowed the $I(E_2)$ curve (Figure 5.4.2D) to be placed on the left side of the $I(E_1)$ curve, we might have restored the steady state. This could have happened if the I neurons had been allowed to inhibit themselves (inhibitory cortical neurons do indeed receive inhibitory synapses from neighboring inhibitory neurons). In the foregoing analysis we have used parallel synchronous updating, which is also deleterious. In this mode of updating, the disturbance takes the form of a sudden synchronous jump in the firing rate (from the E_0 level to the E_1 level). In realistic networks, this increment will occur gradually during the 1-ms time interval between the two steps of the simulation. Thus, parallel asynchronous updating may yield a higher stability. A considerable source of instability in the network is the extra lag required for the effects of the inhibitory neurons to feed back to the excitatory neurons.

Exercise 5.4.1. Assume that inhibition is multiplicative, and draw a qualitative set of graphs showing the firing rate of an inhibitory neuron in the network as a function of the rate of firing of the excitatory neurons that will be consistent with the graphs in Figure 5.4.2.

Exercise 5.4.2. Build up a network for which the I neurons inhibit the other I neurons as well, and try to set up the set of curves (E to E and E to I) that will assure stability.

The graphs in Figure 5.4.2 are not realistic for cortical networks because their slopes are too flat. We shall see (Section 7.1) that at the ongoing rate in the cortex, the threshold is about 25 EPSPs above the average membrane potential. This means that near the equilibrium point, the input–output curve should look like

$$y = kx^{25}$$

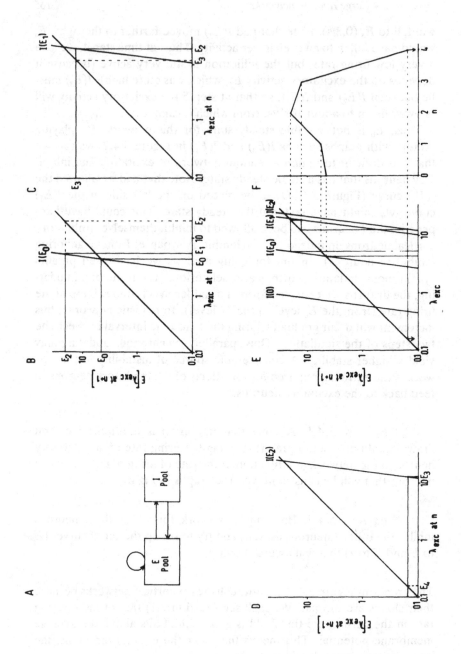

The slope of this curve is

$$y' = 25kx^{24}$$

at the steady-state point $y = x$ (the output rate equals the input rate), and we have

$$y = ky^{25}$$
$$y' = 25ky^{24}$$
$$\therefore$$
$$y' = 25$$

The slope at the steady-state point is twenty-five! That is, even a slight increase in the spontaneous firing rate in the excitatory cells will be amplified twenty-five times before the inhibition has a chance to overcome it. This may be beyond the range of stability that can be attained by inhibitory feedback. This argument will not hold for steady states attained at high firing rates (as were achieved in many of the nets described in Section 5.3 and in Section 5.4.1), because there the threshold is quite close to the average membrane potential, and the slopes near the steady-state firing rates will not be steep.

Conclusion. For a large network of excitatory and inhibitory neurons with small EPSPs it is very difficult, if not impossible, to attain steady ongoing activity at low firing rates. The problem of attaining a low level of ongoing activity is due to the steep slope of the input–output curve of the excitatory population at low firing rates and to the extrasynaptic delay required before the inhibitory cells can counteract random fluctuations in activity. If it is at all possible to attain stable ongoing activity at low levels, it is likely that the range of fluctuations around which the system can be stabilized will be very limited, that the

Figure 5.4.2. Evolution of activity in a large network with excitatory and inhibitory neurons. A: Structure of the network. B–D: Expected firing rate of an excitatory neuron as a function of the averaged firing rate of the excitatory neurons in the network. The graphs use a double logarithmic scale and therefore appear straight, with a slope of T (assumed here to be eight). As the excitation increases, so does the inhibition, and therefore the input–output curves shift. But because of the extrasynaptic delay, they always lag one step behind the excitation. B, C, and D show the evolution of activity after one, two, and three time steps. E: Superposition of B–D. F: Evolution of activity with time. A small disturbance (from E_0 to E_1) is at first amplified strongly, then restrained, but then activity falls off steeply, and the network is turned off.

inhibitory neurons will have to switch from no activity to maximal activity over a narrow range of excitatory firing rates, and that the inhibitory neurons will also have to exert strong inhibitory effects on themselves.

Problem 5.4.1. Analyze rigorously the problem of maintaining stable low levels of activity in a large network of excitatory and inhibitory neurons.

The delicate balance between the internal excitation and inhibition in large networks sheds light on some of the anatomical details of the cortex. The need to switch from no inhibition to fast inhibition explains why only few inhibitory neurons having very strong effects are found in the cortex. The need to have the inhibitory feedback with as short a delay as possible explains why there is no use for extracortical inhibitory feedback (conduction times to such subcortical nuclei and back are too long). This requirement for a short delay also explains why the axons of most inhibitory cortical neurons are thicker than the average cortical axon (fast conduction times), why they distribute their effects mostly in their own vicinity (short conducting distances), and why their postsynaptic targets are concentrated on the cell bodies and the proximal dendrites (shorter electrotonic delays to the soma and faster rise time of the IPSPs).

The synaptic arrangement in which excitatory neurons receive a mixture of excitatory and inhibitory synapses on their somata (and large dendritic trunks), whereas the inhibitory neurons receive excitation only on more remote dendritic sites, also contributes to fast inhibitory feedback and more sluggish positive excitatory feedback. The effective mutual inhibition that is exerted by the inhibitory neurons on each other also seems necessary for attaining stability.

The conjecture that the inhibitory feedback can stabilize the ongoing activity only over a narrow range of firing rates sheds light on the arrangement of the specific thalamic inputs to the cortex. These inputs bring strong excitatory effects to the cortex (particularly to sensory areas). The thalamic inputs spread their terminals in cortical layer IV, where they make synaptic connections with all the neural elements that pass through the layer [White, 1989]. Layer IV is particularly rich with inhibitory neurons. In this manner, the input that increases the activity in the cortical excitatory neurons concomitantly activates the inhibitory neurons. This is a feed-forward activation of the inhibitory neurons that anticipates the increased excitatory activity and alleviates the problems that the extra delay of inhibitory feedback might cause.

The difficulty of stabilizing the cortical activity explains why it is so easy to disturb the cortical activity in pathological ways. Any external interference with the delicate balance of excitation and inhibition can throw the cortex into saturated activity (epilepsy), depress its activity completely (spreading depression), or drive it into oscillations. However, in the cortex, inhibition plays additional roles. It is known that inhibitory mechanisms take part in shaping the response properties of cortical neurons [Sillito et al., 1980] and, presumably, in any other computations conducted by the cortex.

5.4.3 Externally driven ongoing activity

For the netlets described in Section 5.3, we saw that there were only two stable steady states: at no activity at all, or at very high levels of activity. However, when external excitation was added (Figure 5.3.3A), then the quiescent stable steady state was replaced by a steady state with a low level of activity. This is intuitively obvious. If a network receives external excitatory input, then even when no neuron fires at time step n, some neurons may be driven to fire (by the external inputs) at time step $n + 1$. The network's dynamic curve does not start at $(0, 0)$ but at $(0, a)$, $a > 0$. This also holds true for the large network. If the network is subject to external input, this input may hold it partially active at low levels. Thus, it is possible that the stable ongoing cortical activity is due not to the stability of local interactions that maintain it but rather mostly to the excitatory input received by the cortex through the white matter.

A straightforward test for assessing the contribution of cortical input to ongoing cortical activity is to undercut the white matter while sparing the cortical blood supply, which is derived from vessels running on its surface. This type of undercutting was carried out extensively by Burns [1958] and, more recently, by Burns and Webb [1979]. The invariable result of such undercutting was complete shutoff of the neurons in the isolated cortical slab.

Traditionally, that lack of activity was attributed to some kind of "shock" caused by cutting the axons in the white matter. However, in view of the foregoing discussion, these experimental results seem to indicate that the cortex is internally at its quiescent steady state and that the observed ongoing activity is due to the continuous bombardment of excitatory inputs derived from axons that arrive through the white matter.

A similar conclusion was recently published by Rubin and Sompolinsky [1989], who studied the behavior of a fully connected network of excitatory and inhibitory neurons. They rigorously demonstrated that the only way to attain a stable low level of ongoing activity, without the need for fine tuning of the synaptic strengths, was to set the threshold of the excitatory neurons at *negative* values. This seemingly nonphysiological requirement is equivalent to stating that the excitatory neurons receive external excitatory inputs that can cause them to fire even when all the neurons in the network are quiescent.

Recalling that more than 99 percent of the axons passing through the white matter connect one cortical area to another (Chapter 1), we can view the entire cortex as one huge neural network [Katznelson, 1981]. One may well ask how the entire cortex maintains a stable low level of ongoing activity. The arrangement whereby all the inhibition is local, whereas much of the excitation is arriving through the axons of neurons from other cortical regions, assures that the average delay of the inhibitory feedback is much shorter than the average delay of the excitatory feedback. To illustrate this, let us assume that at a given moment there is a slight increase in activity in the excitatory neurons in a certain cortical region. This superfluous excitation will activate other cortical areas after a few milliseconds, and these will feed back excitatory inputs to the original area after a few more milliseconds (If the average distance between two cortical areas is 30 mm and the average diameter of an axon in the white matter is about 1 μm [Tomasch, 1954; Bishop and Smith, 1964; Haug, Kolln, and Rost, 1976], then the conduction velocity in these axons is about $6 \times 1 = 6$ m/s, and the time it takes for excitation to travel back and forth is $2 \cdot 0.03/6 = 10$ ms.) However, the local inhibitory feedback will begin to suppress the additional excitatory activity after only 2 ms.

We note that there are strong indications that certain modes of cortical activity are not generated within the cortex, but are reflections of the activity in the thalamic input to the cortex [Anderson and Anderson, 1974].

To summarize, it appears that in each cortical area, a major portion of the excitation that drives the ongoing activity is derived from activity coming through the white matter. The inhibition is generated locally. This arrangement assures the local stability of the ongoing activity. When the entire cortex is viewed as one huge network, stability is assured by the short delay of the local inhibitory feedback, as compared with the conduction times of excitation through the white matter.

5.4.4 Modes of activity in random networks

The ongoing activity of a random network need not be random. Rather, it can show several features that maintain themselves in a stable manner. The netlet described in Section 5.3 has N neurons, and the firing constellation (the list of firing neurons) at time $t = 1$ depends only on the firing constellation at time t. This netlet has only 2^N different states (= firing constellations); therefore, its activity must be periodic, and the maximal possible period time is 2^N ms. However, Anninos [1972] found that usually the periods are much shorter.

The question of periodicity in random networks has been elegantly analyzed by Sompolinsky, Cristani, and Sommers [1981]. In their model, the firing rate of each neuron depends on the membrane potential according to an S-shaped curve:

$$S = \tanh(gU)$$

where S is the firing rate, U is the membrane potential ($U = 0$ is the resting level), and g is a "gain factor" determining the slope of the relation between the membrane potential and the firing rate. All the neurons in their net are coupled by randomly chosen synaptic strengths that can be positive (excitatory synapses) or negative (inhibitory synapses). The mean synaptic strength is assumed to be zero, and the variance around this mean is σ_J^2/N. They show that when

$$g\sigma_J < 1$$

the network has only one stable steady state, in which all the neurons fire at a fixed rate. When

$$g\sigma_J \gg 1$$

the neurons show fluctuating firing rates, and their behavior is chaotic.

When $g\sigma_J$ is somewhat greater than 1, the behavior (as represented by the autocorrelation function of the firing rates) may be periodic. When $g\sigma_J$ is just slightly above 1, the oscillations are simpole, and as σ_J increases, the oscillations become more and more complex until they become chaotic. The range of values at which oscillations are predominant depends on the size of the system. For 100 neurons,

$$1 < g\sigma_J < 2$$

and for 1,000 neurons

$$1 < g\sigma_J < 1.4$$

Note that the frequency at which the network's activity will oscillate can vary widely when networks with different random connectivities are compared.

This model sheds new light on the behavior of the cortex when subjected to epileptogenic effects. In the cortex, both σ_J and g are large, so in normal conditions one expects to see chaotic behavior. This is expressed by the shape of the autocorrelation function of the electro-encephalogram (EEG) in the awake brain, which falls off exponentially to zero. At this stage, the firing rates of the neurons fluctuate at low levels (in Sompolinsky's model this state is represented by S being negative most of the time). When an epileptogenic drug is applied to the cortex, the activity of all the neurons is greatly increased, and the EEG fluctuation becomes very large but is still chaotic – the so-called tonic phase of the seizure. During this phase, the firing rates of the neurons fluctuate at high levels (S is positive in Sompolinsky's model). As the tonic phase continues, the synapses show fatigue, which means that σ_J decreases gradually. Eventually it becomes so small that $g\sigma_J$ is close to 1, and the seizure takes the form of periodic waves. This is the so-called clonic phase of the seizure. With a further decline of the synaptic strength, $g\sigma_J$ becomes smaller than 1, and the cortex drops to a steady quiescent state.

Periodicities during the clonic phase certainly are affected by processes such as accumulation of potassium and depletion of calcium. However, the possibility of transition from low-level chaotic activity to high-level chaotic activity, then to periodic activity, and finally to complete depression was shown to be a general property of randomly connected neural networks.

Katznelson [1981] considered the entire cortex as one huge two-dimensional network that takes the form of the mantle of a sphere. By assuming that the strength of coupling between cortical regions falls off exponentially with the distance between them, he obtained a structure in which perturbations tended to propagate along the sphere's surface and would return to their source of origin after characteristic times. These properties defined wave periodicities and shapes at which the entire cortex could maintain stable activity. Katznelson was able to show that certain features of the EEG obtained under the effect of anesthesia could be predicted from his model.

Several models describe the short-range connectivities in the cortex by differential equations [Wilson and Cowan, 1973; Ermentrout and Cowan, 1979; Krone et al., 1986]. In all these models a simple activation of a small region sets in motion complex waves of activity in space and time.

To summarize, even completely random networks can generate stable firing patterns (modes of activity). Using models, one can create a whole spectrum of stable modes by a variety of assumptions regarding the connectivity, the distribution of synaptic strengths, and the internal properties of each neuron. The question remains to what degree the modes of activity observed in the cortex are due to the internal structure of the cortex and to what degree these modes are due to pacemakerlike activity derived from deeper nuclei. (See Andersen and Andersson [1974] for a review of physiological perspectives regarding the source of organized activity in the cortex.)

5.5. Recurrent cooperative networks

Thus far we have dealt only with random networks. We have seen that investigation of such networks can shed some light on phenomena observed in the cortex. However, it is quite clear that such networks do not compute anything useful. The preceding chapters dealt with the feasibility of small neuronal circuits that operate on a background of random connectivity and activity. In this section we briefly discuss a class of models used to describe behavior in large networks that carry out useful computations, such behavior being an emergent property of the collective interaction among all the neurons in the net. The description essentially follows the work of Hopfield [1982, 1984; Hopfield and Tank, 1986]. The interested reader is referred to recent monographs by Peretto [1989], Gestzi [1990], and Amit [1989], whose notation is adopted in this section. The reader who is familiar with these networks is advised to turn directly to Section 5.5.3, which discusses relations between recurrent neural networks and the cortex.

5.5.1 The basic model

We deal with a network of N point neurons (as described in Section 5.2) that sum up linearly all the synaptic inputs. If the summed inputs are above threshold, the neuron fires. Time is quantized (i.e., $t = 0, 1, 2, \ldots$). Usually, the time unit is equivalent to one synaptic delay (about 1 ms). For some models this time unit is divided into N (the

number of neurons) substeps. This issue is explained in detail in the
following paragraphs.

Each neuron in the network may be connected synaptically to all the
$N - 1$ other neurons, but usually not to itself. The strength of the
junction from neuron j to neuron i is denoted J_{ij}. It is already scaled to
units of millivolts, so that when neuron j fires, it adds J_{ij} mV to the
membrane potential of neuron i. If neuron j excites neuron i then $J_{ij} > 0$;
if, on the other hand, neuron j inhibits neuron i, then $J_{ij} < 0$. Usually, it
is assumed that a neuron never contacts itself ($J_{ii} = 0$).

The duration of all the synaptic effects is one time step, and the
neurons are refractory for exactly one time step. The membrane potential (U) of a neuron at time $t + 1$ depends only on the firing states of all
the other neurons at time t.

$$U_i(t + 1) = \sum_{j \neq i} \sigma_j(t) \cdot J_{ij} \qquad (5.5.1)$$

The membrane potential is assumed here to be zero at rest. The neuron
fires according to some threshold function that in the simplest cases
looks like a step function:

$$\sigma_i(t) = \begin{cases} 1 & \text{for } U_i(t) > \theta \\ 0 & \text{for } U_i(t) < \theta \end{cases} \qquad (5.5.2)$$

The state of the network at any time t is defined by the list of neurons
that fired at that time: $\sigma(t)$. For example, in a network of four neurons,
$\sigma(5) = (0, 1, 0, 0)$ means that at time step 5, neuron number two fired,
and the other three were quiescent. When starting from some initial step
$\sigma(0)$, equations (5.5.1) and (5.5.2) completely determine the evolution
of activity in the network.

The manner in which the network's state is updated [equations (5.5.1)
and (5.5.2)] may strongly affect the evolution of activity in the network.
There are two principal ways to update the network's state: parallel and
serial. In parallel updating, the state (firing/not firing) of each neuron is
updated by considering the state of its presynaptic inputs at one synaptic
delay earlier. The updating of the state of the netlet described in Section
5.3 was done using this method.

In serial updating, only one neuron is updated at a time. Once updated, it immediately affects the next neuron to be updated. Thus, the
serial updating method does not take into account the synaptic delay in
its physiological sense. The neurons to be updated can be chosen in a

fixed order (*sequential updating*) or picked at random (*random updating*). When considered physiologically, N time steps of serial updating are equivalent to one synaptic delay time.

The last point to be mentioned here is that parallel updating can be done in two forms: synchronous and asynchronous. Synchronous parallel updating means that the state of all the neurons is updated instantaneously at the beginning of each time step, and then for one synaptic delay nothing is changed. Only at the beginning of the next time step is the state of all the neurons updated again. Asynchronous parallel updating means that the synaptic delay is broken into many subsidiary time steps. At each substep the state of all the neurons that are not refractory is updated based on the network's state at one synaptic delay earlier. Although this is the best way to mimic a network of real neurons, asynchronous parallel updating is seldom used for neural network modeling.

To summarize, there are four types of modes of updating the network's state: sequential, random, synchronous parallel, and asynchronous parallel. In sequential and random updating, time advances in $1/N$ of a synaptic delay, and the neurons have immediate effects on each other. In parallel updating, the effects appear only after a synaptic delay. In synchronous parallel updating, time advances in one synaptic delay, whereas in asynchronous parallel updating it advances in a fraction of a synaptic delay. The latter mode is the most physiological, but the least frequently used for recurrent networks.

5.5.2 The behavior of the network

A major feature that one can look for is the presence of stable firing constellations; that is, constellations for which $\sigma(t) = \sigma(t + 1)$. Once a network reaches such a state, it stays there forever. Note that these stable steady states are not due to interactions in small groups of neurons, but rather constitute an emergent collective property of the cooperative interactions of all the neurons within the network. For that reason, the name "recurrent cooperative networks" is chosen here. Other names to be found in the literature are "attractor neural networks," "associative networks," "Little and Hopfield networks," and "spin glass networks."

The notion that the brain does not compute by means of specific circuits made of a few neurons, but rather by interactions within large masses of neurons, is not new. Failing to demonstrate specific memory deficits caused by localized lesions to the brain, Lashley [1929] formulated his theory of mass action within the nervous system. Cragg and Temperly [1955] suggested an analogy between the cortex and a system of small magnets interacting with each other and creating domains of alignment and disalignment. Caianiello [1961] set up the equations for solving the behavior of a large number of neurons interacting with each other. John [1972] presented experimental evidence for the need to consider the interactions between large masses of neurons. Little [1974] described the entire brain as an Ising system of spins and worked out the conditions that would ensure the existence of persistent states within the brain. However, the major step toward solving the cooperative behavior of neural networks was made by Hopfield [1982], who added the two following assumptions: (1) The interactions between neurons are symmetric:

$$J_{ij} = J_{ji} \tag{5.5.3}$$

that is, the matrix of connection $\mathbf{J}$ is a symmetric matrix. (2) Updating can be carried out in serial or random fashion. When these two assumptions hold true, the network is similar to a system of spins (little magnets) that force each other to align or disalign, and in which, by the law of "action and reaction," the force that spin j exerts on spin i is the same as the force exerted by spin i on spin j (physicists call such a system of magnets the "Ising model"). Naturally, in a system of such spins there is no "synaptic delay," and therefore sequential (or random) updating is adequate. The major advantage of describing a neural network as a physical system is that such physical systems seek states σ with local minimal energy. This has been used by Hopfield and his followers in elegant investigations of the properties of such networks.

To complete the analogy between the neural network and the system of spins, the following symbolic conversions are made:

$$S_i = 2\sigma_i - 1 \tag{5.5.4}$$

$$T_i = 2\theta_i - \sum_{i \neq j} J_{ij}$$

The change from σ to S changes the notation of the neuron's state (not firing/firing) from $(0, 1)$ to $(-1, 1)$, which corresponds to the spins being in either one direction or the other. The change of threshold from θ to T is then required in order to offset the negative effects that nonfiring neurons ($S_j = -1$) exert on the membrane potential. We note that as long as the junction strengths (J_{ij}) are fixed, the conversions of equations (5.5.4) are simply a matter of convenience. However, in neural network models that learn (by changing the junction strengths), the new convention implies that as the network learns, the new thresholds (T_i) must also change.

Using this new notation, we define a discontent function

$$D[S(t)] = -\frac{1}{2} \sum_{j,i} S_i(t) J_{ij} S_j(t) + \sum_i S_i(t) T_i \qquad (5.5.5)$$

which is to be understood as follows: The "discontent" of the network is the sum of the discontent of all the synapses (the first term) and the discontent of the neurons (the second term). If we note the minus sign before the first sum, we see that the discontent is increased when $-S_i J_{ij} S_j$ is positive. This term is interpreted as the local discontent of a synapse, because when J_{ij} is positive (the junction is excitatory), the local discontent is positive when the postsynaptic neuron (i) and the presynaptic neuron (j) do not act in consonance [i.e., either the postsynaptic neuron is firing ($S_i = 1$) when the presynaptic neuron did not ($S_j = -1$), or vice versa]. Thus, an excitatory synapse between neurons j and i is said to be discontented when the presynaptic and postsynaptic neurons show opposite signs of activity. An inhibitory junction ($J_{ij} < 0$) is discontented when the presynaptic and postsynaptic neurons act coherently. The local discontent is proportional to the junction strength J_{ij}, so that if neuron j fires and neuron i does not, and the synapse between them is strongly excitatory, then that synapse contributes a large value to the discontent. To the discontent of the synapses we add the discontent of the neurons [the second term in equation (5.5.5)]; this is large when a neuron with a high (positive) threshold fires, or when a neuron with a low (negative) threshold does not fire.

When the state of a single neuron is updated, it cannot increase the total discontent, as is shown in the following discussion: Suppose the state of neuron 1 is to be updated. Shortly before updating, the sum of its discontent and the discontent of all the synapses it receives is

$$-\frac{1}{2} S_1 \left\{ \sum_j J_{1j} S_j - T_1 \right\} \qquad (5.5.6)$$

The sum $\Sigma J_{1j} S_j$ is the membrane potential of neuron 1 [equation (5.5.1)]. Therefore, if $\Sigma J_{1j} S_j > T$, then $S_1(t+1) = 1$, whereas if $\Sigma J_{1j} S_j < T$, then $S_1(t+1) = -1$. Thus, at time $t+1$,

$$-\frac{1}{2} S_1(t+1) \left\{ \sum_j J_{1j} S_j - T_1 \right\} < 0 \qquad (5.5.7)$$

If the state of neuron 1 does not change, there is no change in the total discontent, whereas if the state of neuron 1 changes, it reduces its own discontent [because after the change, the left-hand side of equation (5.5.7) is negative, and that means that before updating, equation (5.5.6) was positive]. What, then, will be the effect of the change in the state of neuron 1 on the discontent of its postsynaptic targets? This contribution is given by

$$-\frac{1}{2}\sum_j S_j(t)J_{j1}S_1(t+1) \qquad (5.5.8)$$

which by virtue of the symmetry of the interneuronal coupling ($J_{1j} = J_{j1}$) can be rewritten as

$$-\frac{1}{2}S_1(t+1)\sum_j S_j(t)J_{1j} \qquad (5.5.9)$$

Expression (5.5.9) is identical with the left-hand side of expression (5.5.7), and therefore we conclude that the change in the state of neuron 1 reduces both its own discontent and its contribution to the discontent of all the other synapses.

For this reason, when a network with symmetric coupling is serially updated, the network seeks states with minimal discontent.

The "discontent function" was called "energy" by Hopfield [1982]; a similar function that takes into consideration the existence of (termal) noise was called "free energy" by Amit, Gutfreund, and Sompolinsky [1987a]. Somtimes it is called "Hamiltonian," "Lyapunov function," or a cost function. The name "discontent" is adopted here because it describes the state of the individual junctions and neurons in a physiological sense.

The discontent function has the property that for a symmetric **J** when the network is left on its own, it will behave in such a way as to minimize the discontent. That is,

$$D[\mathbf{S}(t+1)] \le D[\mathbf{S}(t)]$$

just as physical systems behave so as to decrease their free energy.

One can imagine the spiking constellation of the system $\mathbf{S}(t)$ (i.e., the list of spiking states of the N neurons) as a point in N-dimensional space. The dynamics of the network can be represented by movements of this representing point in the states' space. For each possible constellation,

Table 5.5.1. *The discontent of a two-neuron "network"*

State	Discontent
$-1,-1$	-2
$-1,\ 1$	1
$1,-1$	1
$1,\ 1$	0

one can compute its discontent D. Constellations at which D assumes minimal values locally are stable.

Example 5.5.1. We have a "network" of two neurons that excite each other, with a strength of 1, the threshold of which is 0.5. What are the discontent values for this system?

This network has four possible spiking states: $(-1, -1)$, $(-1, 1)$, $(1, -1)$, and $(1, 1)$, for each of which we can compute the discontent. For instance,

$$D[(-1, -1)] = -\tfrac{1}{2}[(-1)1(-1) + (-1)1(-1)] + [(-1)0.5 + (-1)0.5]$$
$$= -2$$

In this manner we can tabulate (Table 5.5.1) the discontent for all the possible states. We see that the state in which all the neurons are quiescent $(-1, -1)$ is stable. The $(1, 1)$ state (in which all neurons fire) has a higher discontent, but it can pass to the state $(-1, -1)$ only through $(-1, 1)$ or $(1, -1)$, both of which have even higher discontent (remember that we update one neuron at a time). Thus, whereas $(-1, -1)$ represents the global minimal discontent, the state $(1, 1)$ represents a local minimum that is stable as well.

The two stable states $(-1, -1)$ and $(1, 1)$ are in accordance with our conclusions in Sections 5.3 and 5.4 that an all-excitatory network with no noise can have only two stable states (when all the neurons fire or all of them are quiet).

We can use this simple example to get the sense of the difference between parallel updating and sequential updating. When updating is parallel, after each step (synaptic delay) we observe all the neurons and update their states according to the state of the net in the previous step. For example, we obtain the pattern shown in Figure 5.5.1, in which each

A   B

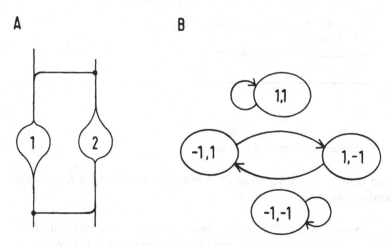

Figure 5.5.1. Behavior of a network of two neurons when the updating is parallel. A: The network. B: The states and the transitions between them.

possible spiking state is described as a node, and the transitions are denoted by arrows. Spiking state (1, 1) is stable because each of the two neurons receives an excitation of 1, and its threshold is 0.5. Spiking state (1, −1) changes to (−1, 1) because neuron 1 receives negative input and will not fire, whereas neuron 2 receives one unit of excitation from neuron 1 and fires, and so forth. We see that the system either is in one of its stable states or oscillates between its two unstable states.

Sequential (or random) updating prevents this oscillation between states, as shown in Figure 5.5.2. Again we note that (−1, −1) and (1, 1) are stable states, whereas (−1, 1) and (1, −1) are not. If the network starts from (−1, 1) and the next neuron to be updated is 1, then it moves to (1, 1), because neuron 1 receives one unit of excitation from neuron 2. Once the network arrives at (1, 1) it will remain there forever. If, on the other hand, the network starts again from (−1, 1) and neuron 2 is to be updated, the network moves to (−1, −1), because neuron 2 receives negative input from neuron 1.

When we deal with more complex networks, the situations become much more complicated, as can be seen in the following exercise.

Exercise 5.5.1. Consider the network shown in Figure 5.5.3, in which the strength of the "dark" synapses is +1, and that of the "light"

A B

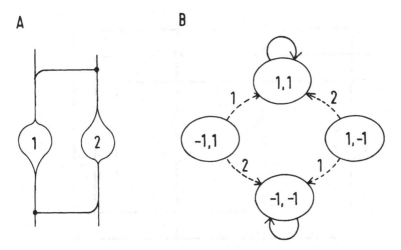

Figure 5.5.2. Behavior of a network of two neurons when updating is sequential. A: The network. B: The states and the transitions between them. The transitions may depend on which neuron is updated. In these cases the transitions are marked by dashed lines and the name of the neuron that is updated appears near the arrow.

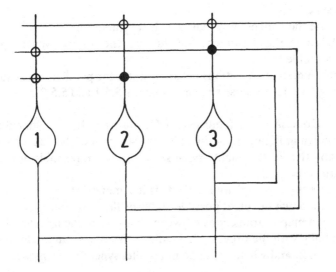

Figure 5.5.3. A recurrent network of three neurons.

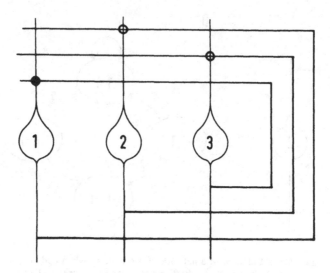

Figure 5.5.4. A recurrent network of three neurons.

synapses is -1. The thresholds of neurons 1, 2, and 3 are -1, 0.5, and 0.5, respectively.

1. Write the connectivity matrix $\mathbf{J}$. Is it symmetric?

2. Tabulate the discontent function for the matrix. Which are the putative stable states?

3. Plot a map of transitions between states for parallel and sequential updating (similar to those shown in Figures 5.5.1 and 5.5.2).

Exercise 5.5.2. (courtesy of H. Gutfreund). Consider the network shown in Figure 5.5.4, in which the strength of the "dark" synapse is $+1$ and that of the "light" synapse is -1. The thresholds of the neurons are zero.

1. Write the connectivity matrix $\mathbf{J}$. Is it symmetric?

2. Write down the eight possible states of the network.

3. Plot a map of transitions between states when the updating is done sequentially with the order 1, 2, 3, when it is done sequentially with the order 1, 3, 2, and when it is done in parallel synchronous mode.

Solving these exercises shows again that for parallel updating we may have stable states or limit cycles, whereas for sequential updating when the connectivity matrix is symmetric, we have only stable states. Around each stable state we can set a region of states from which the state of the net will flow into that stable state. Several states are on saddle points, so that they may flow to one stable state or to another, according to the updating order.

The behavior of a recurrent cooperative network with a symmetric connectivity matrix and sequential updating can be described as follows: All the possible states of the system span an N-dimensional space. The behavior of the network (its change from state to state) can be described as moves in this state space. Over the state space the discontent function defines a landscape of hills and valleys. The network state changes so as always to go downhill on this discontent landscape. Certain states lie at the bottoms of such valleys. These states are said to attract neighboring states, because when the network starts from a neighboring state and it is updated serially, the network's state moves toward one of these states of minimal discontent. The region in the state space from which the network states will flow to a given attracting state is the basin of attraction of this state.

The existence of attracting states surrounded by basins of attraction suggests the use of neural networks for classifying information. The state of the neurons defines an N-bit word. The attracting states are considered as prototypes (denoted by ξ) of a class of states. All the states located in the basin of attraction of a prototype constitute the class of states associated with this prototype. If we force the network at time zero to start from a given state $S(0)$, its state will change until it arrives at a prototype. Simulations of such networks with several hundred neurons show that when the network is started with a firing pattern that considerably overlaps one of the prototypes, the network's state converges to the prototype within a few time steps. A recurrent network that can classify states as described earlier often is referred to as an "associative network" (because it associates initial states with a few prototypes) or "content addressable memory" (because it can reconstruct a whole pattern from a partial version of the pattern).

We note that this brings us to a very general conclusion: If in a fully connected network of neuronlike elements the connections are symmetric, then when left alone, the network will reach a fixed point at which some of the neurons will fire at maximal rates, whereas others will remain quiescent. One cannot pinpoint a specific partial circuit that forces the neurons to fire (or to shut off); rather, this behavior of the network is the result of the cooperative effects of all the neurons on each other.

The detailed constellation of such a fixed point (i.e., which neurons are firing and which are quiet) depends on the values of the junctions J_{ij} between the neurons. These can be programmed to produce any desired fixed point. We can prescribe a fixed point at which a given group of neurons will fire (and the rest will be quiet) by introducing strong excitatory synapses between every pair of neurons in the firing group, strong inhibitory synapses between every pair of neurons of which one is in the firing group and the other is in the quiet group, and excitatory synapses between pairs of neurons from the quiet group. We note that by this prescription we have programmed not only the desired state but also its complement (in which all the neurons we wished to be quiet will fire at the maximal rate, and all the others will stay quiet) as fixed points.

How do we proceed if we wish to embed a whole set of prototypical states ($\xi^{(1)}$, $\xi^{(2)}$, . . . , $\xi^{(P)}$? This can be done by setting the synaptic strengths such that

$$J_{ij} = \frac{1}{N} \sum_{\mu=1}^{P} \xi_i^{(\mu)} \xi_j^{(\mu)} \tag{5.5.10}$$

where P is the number of prototypes, and $\xi_i^{(\mu)}$ is the state of neuron i in the μth prototype. That is, if in the kth prototype both neuron i and neuron j have consonant activities (either both are on, or both are off), we should make the junction between them slightly ($1/N$) more excitatory, whereas if the two neurons have dissonant activities, the junction must be slightly more inhibitory. The final strength of each synapse is the sum of the contributions of the presynaptic–postsynaptic coherence over all the prototypes. This mode of "programming" the network to take a set of desired attractors is called "Hebb's rule" in honor of D. O. Hebb, who suggested that synapses in the brain become stronger if there is a correlation between the presynaptic and postsynaptic activities [Hebb, 1949]. Note that the programming rule of equation (5.5.10) dictates symmetric synaptic strengths:

$$J_{ij} = J_{ji}$$

We note that Hebb originally spoke of strengthening synapses in which the presynaptic activity slightly preceded the postsynaptic activity. Thereby, Hebb's cell assemblies produced a sequence of activities at the recalled memory, not a group of neurons that coactivated each other in a stable fashion.

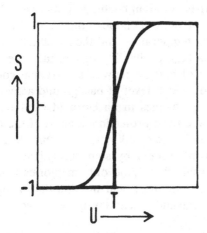

Figure 5.5.5. A sharp threshold function and a gradual function.

One can preprogram the network according to equation (5.5.10), or one can have a learning stage in which the patterns $\xi^{(\mu)}$ are imposed one at a time on the network and the synaptic strengths are modulated according to Hebb's rule. An attractive feature of Hebb's rule is that it is local (i.e., the strength of each synapse is modulated only according to what happens in the presynaptic and postsynaptic neurons).

When embedding multiple memories in a network, one faces the problem that in addition to the main valleys associated with the embedded memories, one generates multiple shallow valleys in the discontent terrain. It is possible to prevent the network from being trapped in one of these shallow valleys by making the transition from the firing state to the nonfiring state gradual, as shown in Figure 5.5.5.

The gradual, S-shaped threshold function is interpreted differently in different models. In the work of Little [1974] and Amit et al. [1985a], for instance, the curve represents the probability of firing as a function of membrane potential. When looked at in this way, the gradual threshold function is equivalent to the effect of random fluctuation on the state of the small magnets in the Ising model [Glauber, 1963]. One often sees the term "temperature" used to describe the degree of graduality of the threshold function. The step function is obtained at zero temperature. The effect of having the network operate at nonzero temperature is to allow its state to have some degree of random fluctuations around the stable states. When the network's state is caught in a spurious, shallow valley, it is "shaken out" by these random fluctuations and has a higher probability of flowing toward the bottom of one of the major valleys. By starting to work at some finite temperature and then gradually cooling down the network, one has a high probability of reaching one of the major stable states. In terms of cortical networks, "gradual cooling" means starting from a state of a high level of background activity and gradually lowering it. Periodic changes in background excitation are known to occur by way of the diffuse projection systems of the cortex (Section 1.3.3), and they result in the well-known α rhythm (8–12/s) in the EEG. This rhythm was interpreted by some investigators as an expression of neural mechanisms that "read out" memories from the cortex [Crick, 1984]. However, one must bear in mind that α rhythms are prominent in the drowsy state and absent during the active, wakeful state of the brain.

Another interpretation of the gradual, S-shaped threshold curve [Hopfield, 1984] is that it represents the firing rate of the neuron. The neuron is at state 0 when quiescent and at state 1 when firing at its maximal possible rate. The transition from low to high firing rate occurs rapidly near the threshold. When looked at in this way, the network's behavior is seen to be governed by a set of differential equations. The notion of a "time step" must be replaced by the "characteristic time," which is approximately the duration of the integration of presynaptic events by postsynaptic neurons (i.e., 5–10 ms). "Low" firing rates are defined as below 100/s, and "high" firing rates must be several hundred spikes per second.

Another problem that arises when embedding multiple memories in a network is the appearance of stable states that look like linear combinations of the embedded states. In neural network modeling, this can be reduced by specifying that the embedded states be uncorrelated with each other. That is,

$$\frac{1}{N} \sum_{i=1}^{N} \xi_i^{(\mu)} \cdot \xi_i^{(\nu)} = \left\{ \begin{array}{ll} 1 & \text{if } \nu = \mu \\ <1/\sqrt{N} & \text{otherwise} \end{array} \right. \tag{5.5.11}$$

where, again, $\xi_i^{(\mu)}$ is the state of neuron i in the μth embedded memory.

Exercise 5.5.3. For a network of eight neurons, write a set of state vectors that are uncorrelated with each other, with the constraint that in each state half the neurons must be firing and half must be quiescent.

However, even when embedding only uncorrelated prototypes with Hebb's rule, for every embedded prototype $\xi^{(\mu)}$ the state $-\xi^{(\mu)}$ becomes just as stable. In addition, mixtures of odd numbers of prototypes also become stable (albeit with higher discontent).

There is a considerable theoretical effort under way to analyze the memory capacity of recurrent cooperative networks. When one attempts to embed more and more prototypes into a recurrent network, one reaches a critical number above which the system "falls apart." According to Amit et al. [1985b], in a network of N neurons with symmetric connectivity and uncorrelated prototypes, as long as the number of embedded prototypes (P) is smaller than $0.14N$, one obtains recall with a low probability of errors. As soon as P exceeds $0.144N$, the probability of error increases very rapidly, and the network becomes useless [Nadal et al., 1986]. Note that the critical value of 14 percent is valid for large recurrent networks with uncorrelated prototypes and with synaptic strengths that are programmed according to Hebb's rule. Programming rules that allow for embedding many more stable (but correlated) states have been developed [Kohonen, 1984; Palm, 1986a; Gardner, 1988].

5.5.3 Cooperative networks and the cortex

Among neurobiologists, one often hears emotional critiques of the applicability of the recurrent cooperative neural network to the brain. Much of the criticism relates to the basic model described in the preceding two sections. Theoreticians stress that although the features of this basic model are required to provide for an easy analytical solution of its behavior, many of them can be relaxed. Indeed, models with many variations have been developed without qualitatively affecting the behavior of the cooperative networks. Considerable effort has also been made to accommodate the model to neurobiology.

Studies of the basic cooperative network model and its variants have provided insight into cortical structure and function. This section evaluates some of the criticism, as well as some of the favorable points. The reader is also referred to a recent summary of this issue by Toulouse [1989].

1. Excitatory synapses are distributed along the dendritic tree. Modeling the neuron as a point neuron (Section 5.2) eliminates the nonlinear interactions that occur along the dendritic tree. This critique is applicable to most of the models described in this chapter, not solely to the cooperative networks. When excitatory synaptic effects are generated along a dendrite, they do not add up linearly. This is so because EPSPs generated on a remote dendrite generate currents flowing through the dendritic tree toward the soma, where the synaptic currents change the membrane potential; on their way these currents may be shunted by other EPSPs (and by IPSPs), which will increase the membrane conductance of the proximal dendrites.

The somatic depolarizing effect of a single EPSP coming from a remote dendrite, when there are no other inputs to the postsynaptic neuron, probably is severalfold greater than its effect on the background of ongoing spontaneous cortical activity. However, one must remember that this comparison is not fair. In the cortex, a neuron receives input from about 100 neurons every millisecond; on this background, when twenty-five additional EPSPs are provided, it will fire. The question of nonlinear summation must be related to the marginal depolarizing effect that an EPSP has when arriving at the 101st EPSP, as compared with its marginal effect when arriving at the 126th EPSP. If all the other synaptic potentials arrived at random locations, the difference might be about 20 percent, which would have only a small effect on the "recalled" memories.

Thus, if the synapses are randomly distributed along the dendritic tree, their nonlinear summation will not disturb the basic behavior of a cooperative neural network. In the literature, one can find many models that do not assume a random distribution, but rather are based on strong nonlinear interactions between adjacent synapses. These models implicitly assume that the targets of various presynaptic sources are prescribed very accurately. The approach adopted here is that deliberations about such models must be deferred until there is experimental evidence to show that growth processes actually exist that can indeed direct pairs (or more) of synapses from particular neurons to be located in close proximity.

2. *Shunting inhibition.* This form of nonlinear summation is certainly applicable to all models of neural networks. The main effect of the inhibitory synapses is to shunt the excitatory synaptic currents, not to add hyperpolarization to the somatic membrane potential.

When stated in this fashion, the depolarization at the soma of neorun i (U_i) is given by the total synaptic current arriving at the soma (Q_i) times the "input resistance" (R_i) of the soma and the proximal dendritic trunks:

$$U_i = R_i \cdot Q_i \qquad\qquad (5.5.12)$$

Taking into consideration the synaptic arrangement of cortical neurons (Section 1.2) that preferentially receive high concentrations of inhibitory synapses on the axon hillock, the soma, and the proximal dendritic trunks, we can state that the "input conductance" (the reciprocal of input resistance) is the sum of the resting conductance (g_i) and all the shunting conductances produced by the inhibitory synapses:

$$\frac{1}{R_i} = g_i + \sum_{j \epsilon I_i} J_{ij} \sigma_j \tag{5.5.13}$$

where I_i is the set of all neurons that make inhibitory synaptic contacts with the soma and proximal dendrites of neuron i. From this perspective, the J_{ij} values are not hyperpolarizations, but rather the shunting conductances contributed by the inhibitory synapses. (Note that we revert here to the σ notation, which is 1 if the neuron fired, and 0 if it did not.)

The total depolarizing current that arrives at the soma can be approximated by adding up all the depolarizing currents withdrawn by the excitatory synapses from the soma:

$$Q_i = \sum_{j \epsilon E_i} J_{ij} \sigma_j \tag{5.5.14}$$

where E_i is the set of all neurons that make excitatory synapses on neuron i. In this sum, J_{ij} is the current that the synapse from neuron j withdraws from the *soma* of neuron i.

Combining equations (5.5.12), (5.5.13), and (5.5.14), we get

$$U_i = \frac{1}{g_i + \sum_{j \epsilon I_i} J_{ij} S_j} \cdot \sum_{j \epsilon E_i} J_{ij} S_j \tag{5.5.15}$$

When expressed in this form, the effects of the inhibitory synapses ($j \epsilon I_i$) and of the excitatory synapses ($j \epsilon E_i$) interacts in a nonadditive form. However, as was shown to me by H. Sompolinsky, as far as firing with a clear-cut threshold (θ) is concerned, equation (5.5.15) can be arranged to a form in which the effects will be additive.

Exercise 5.5.4. A neuron fires whenever $U_i = \theta$. Show that if U_i is given by equation (5.5.15); then the excitatory and inhibitory effects are additive. Interpret the terms

$$\theta g_i \quad \text{and} \quad \sum_{j \in I_i} \theta \cdot J_{ij} \cdot S_j$$

(which you will obtain) in physiological terms.

Thus, for neurons with a steep threshold function for which the shunting inhibition is concentrated around the soma and the proximal dendritic trunks, the additivity of all synaptic effects is a fair approximation.

3. The synaptic connectivity is unnatural. In the simplest cooperative neural network described here, each neuron can generate both inhibitory and excitatory synapses (for a given j there may be i's for which $J_{ij} > 0$ and i's for which $J_{ij} < 0$), which is not likely to occur in the cortex. Furthermore, the symmetry ($J_{ij} = J_{ji}$ does not seem to hold true in nature. These requirements have been relaxed in simulations [Shinomoto, 1987] and analytically [Derrida, Gardner, and Zippelius, 1987] without major effects on the operation of the network as an associative memory. Essentially, such relaxations were made by starting with a standard cooperative network in which several prototypical memories were embedded, and then deleting synapses from the network at random, or according to some rule (e.g., if neuron j is to be excitatory, all its negative J_{ij} values are converted to zero). Cooperative networks are immune to such diffuse damage.

4. Effects of synaptic noise. A cooperative network, when used as an associative memory, is immune to loss of synapses, as described earlier. It is also immune to synaptic noise. Thus, one can clip the synaptic strength to just two values (-1 and 1) and maintain the network's ability to classify memories according to prototypes. We should note, however, that cooperative networks that solve optimization problems are very sensitive to such clipping, or to synaptic damage (I thank P. Peretto for drawing my attention to this point).

5. Pulselike postsynaptic potentials. Although the quantization of time in which the postsynaptic effect lasts only a single time step is convenient for solving the network's behavior analytically, it is not physiological. However, that requirement was relaxed by Buhmann and Schulten [1987], without much effect on the behavior of the network. Note that in the version advanced by Hopfield [1984], where the output of the neuron is not of the yes/no type, but is considered to be a continuous variable (i.e., firing rate), the effects may last for the duration of the postsynaptic potential (5–10 ms).

6. *Nonphysiological firing rates.* When in the stable state, some of the neurons of a cooperative network fire at a maximal rate (i.e., 500–1000/s), whereas others are completely quiet. In the standard associative memory models, half the neurons are in the "on" state (they fire maximally), and half are in the "off" state. This situation corresponds to an average firing rate of several hundred spikes per second for every neuron, which is way beyond what is observed in the cortex. However, models have been proposed in which the average firing rate can be as low as desired [Amit et al., 1987b]. Some aspects of behavior of networks were improved by this modification. In early models with low firing rates there were always a few neurons that fired at a very high rate, and all others were quiet. In primary (sensory and motor) cortical areas, some neurons may fire at such high rates. However, in associative areas, the firing rates of "responding" neurons are in the range of 10–30/s. In these areas, only rarely will a neuron fire at a rate above 70/s. In more recent adaptations [Treves and Amit, 1989; Rubin and Sompolinsky, 1989; Buhmann, 1989], even the neurons that are "on" may fire at physiological rates.

7. *No synaptic delay.* Many of the early theoretical results regarding the behavior of recurrent cooperative networks were obtained when serial updating was employed (Section 5.5.1). In this mode, every neuron exhibits essentially immediate effects on all the others. A synaptic delay of approximately 1 ms is certainly inconsistent with this mode of updating. Parallel updating seems more appropriate for modeling networks with synaptic delays. The behavior of networks with parallel updating is less orderly, as was illustrated in the examples of Section 5.5.2. Some amelioration might be achieved if one used the analog output model with a characteristic time of 5–10 ms and a synaptic delay of 1 ms.

8. *Overloading.* Some weak properties of cooperative networks are apparent when the general behavior of the model is considered (see items 9 and 10 in this listing). When a cooperative network (in which Hebb's rule is used to embed multiple memories) is slightly overloaded, it falls apart completely. That is, so many spurious prototypes appear that recall of *all* memories becomes very poor. By contrast, psychologi-

cal experience seems to indicate that slight overloading affects some memories, but leaves most of them intact.

9. Integration with other areas. The cooperative network exhibits two stages of activity. In the first stage the network's state is changing, flowing down the discontent gradient. In the second stage it remains around the state of minimal discontent. Only while in the second stage does the activity of the neurons contain useful information for the rest of the brain. It is not known how other brain regions, receiving inputs from a cooperative network, distinguish between the searching stage and the resolution stage.

10. Coactivation. The one property of recurrent cooperative networks that is most amenable to experimental verification is as follows: When the network reaches resolution (a state of minimal discontent), some of its neurons must fire at a high rate, and the others must remain quiet. This is an essential property of the cooperative network. This property leads to the prediction that when one neuron fires a high-frequency bursts, its neighbors are likely either to be quiet or also fire a high-frequency burst. This prediction can be tested experimentally; however, to date there have been no published data of this sort. Preliminary studies in our laboratory have failed to provide evidence for such coactivation [Abeles, Vaadia, and Bergman, 1990].

11. Immunity to noise. Cooperative neural networks have several attractive features. Their immunity to damage and to synaptic noise has already been mentioned. In a large cooperative network, a single neuron contributes only a trifle to the total synaptic effect on all the other neurons. Therefore, the behavior of the entire network is extremely robust. Deletion of neurons and of synapses, or adding synaptic noise, or any other kind of noise, will have little effect on the tendency of the network to seek states of minimal discontent and little effect on the nature of these states. In this respect, the behavior of recurrent cooperative networks is similar to that of a living organism, where scattered neuronal death, or the random nature of transmitter release by the synapses, will have little effect on overall observable behavior.

12. Structure. The structure of cooperative networks, in which every neuron is connected extensively to many other neurons, is very similar to what is found in the cortex. The result of a computation carried out by a recurrent cooperative network is defined by the state of the network; that is, one must know for each neuron in the network whether or not it fired. That means that the axons of all the network's members must come out of the network to provide the information about the network's state. The cortex, in which the majority of neurons (the pyramidal neurons) provide output through the white matter, meets this requirement.

13. Collective emergent properties. The models of cooperative networks have brought home the idea that there may be neural structures whose behaviors cannot be understood by sampling just a few neurons. It is the collective massive interaction among all the neurons in the network that determines its behavior. For that class of computation, working out simple anatomical circuits, or measuring the activity of just a few neurons, makes no sense.

In conclusion, models of recurrent cooperative networks are rich and fascinating in their own right. There is no a priori theoretical ground for rejecting their applicability to the cortex. Whether or not the cortex computes by seeking stable states of minimal discontent remains a question for experimental verification.

5.6 Perceptrons

This section gives a brief and superficial exposition of perceptrons and of multilayered perceptrons. Most of the ideas presented here are derived from what I have learned from E. Domany, to whom I am most grateful. The reader who wishes to acquire a deeper understanding and become more familiar with current trends in this field is advised to consult the two volumes by Rumelhart and McClelland [1986]. The reader who is familiar with the basic concepts of perceptrons is advised to turn directly to Section 5.6.3, in which relations between multilayered perceptrons and the cortex are discussed.

5.6.1 Simple perceptrons

The term "perceptron" was introduced in the 1950s to designate a simple mechanism to achieve "perception." In terms of what has been developed so far in this chapter, a perceptron is a point neuron (Section 5.2), that is, a neuron that adds all the excitatory and inhibitory effects linearly and fires when their sum exceeds the threshold. The perceptron

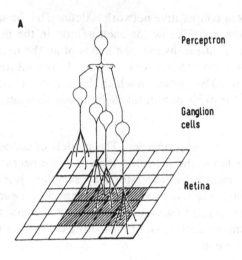

Figure 5.6.1. The perceptron. A: The basic idea: A receptor sheet (retina) is connected to many local analyzers (ganglion cells), all of which are connected to the perceptron. The receptive fields of the ganglion cells, and the strength of the connections from the ganglion cells to the perceptron, are adjusted so that the perceptron can detect whether or not the stimulus presented on the receptor sheet has some global feature. B: The receptive fields of ganglion cells required for making a perceptron that will fire only if a set of disconnected rectangles is presented.

is unique in that its inputs are composed of "sensory information" coming from analyzers, each of which can detect the existence of a simple input pattern in a small receptive field. The purpose of the perceptron is to make judgments regarding global features of the input.

Example 5.6.1 (based on Dewdney [1984]). A picture is projected onto a "retina" made of $n \times n$ photoreceptors. Behind the retina is a large number of "ganglion cells," each of which looks on a small region, and the outputs of all these ganglion cells converge on the perceptron. Figure 5.6.1A describes this situation. In this example, every ganglion cell is affected by a small receptive field of 2×2 photoreceptors.

Let us suppose we wish to construct a perceptron that will fire only when a rectangle (or a set of disconnected rectangles) is projected on the retina. For that purpose we need a set of six ganglion cells looking at each 2×2 region. Each of these ganglion cells is tuned to fire when it "sees" one of the patterns of Figure 5.6.1B in its receptive field. For an $n \times n$ retina we need $6(n - 1)^2$ such ganglion cells. All these ganglion cells converge on the perceptron with a synaptic strength of -1 (they inhibit the perceptron). If the image on the retina is indeed a rectangle, none of the ganglion cells are excited. If the image on the retina is anything else (including rectangles that touch each other), then at least one of the $6(n - 1)^2$ ganglion cells will fire. If we set the threshold of the perceptron to -0.5, it will fire only when the image on the retina is a rectangle or a set of disconnected rectangles (or nothing).

The notion that a simple structure such as the perceptron could detect global features raised great hopes in the research community working with artificial intelligence (AI) in the 1950s, particularly because the perceptron was accompanied by a *learning rule* and a *convergence theorem*.

The perceptron learning rule is as follows: Start with an arbitrary set of synaptic strengths, present an arbitrary shape, and record the perceptron's output. If it is correct, do nothing. If it is wrong, modify the synaptic strength as follows: If the perceptron failed to fire when it should have, slightly increase the strength of the synapses of all the ganglion cells that fired. If it fired when it should not have, slightly reduce the strength of the synapses from all the ganglion cells that fired.

The convergence theorem states that if there is any set of synaptic strengths (J_j) for which the perceptron will function as desired, then by applying the learning rule, one will, after a finite number of iterations, obtain a perceptron that also operates as desired.

These features raised hopes of building simple "seeing" machines that would be able to recognize objects regardless of their position, size, or orientation in space (see Rosenblatt [1962] for a full exposition of these views). Those hopes vanished when Minsky and Papert published their book *Perceptrons* [1969], in which they rigorously demonstrated that perceptrons are quite limited in their ability to extract global features from local information.

5.6.2 Multilayered perceptrons

The scientific community overreacted to Minsky and Papert's work. The hopes were so high, and the efforts toward constructing

Table 5.6.1. *Contrast detection in a two-receptor retina*

Receptor		
1	2	Perceptron output
0	0	0
0	1	1
1	0	1
1	1	0

perceptrons with useful capacities so intense, that the severe limitations exposed by Minsky and Papert resulted in almost complete abandonment of this line of thought. In recent years, researchers have realized that if, instead of demanding a one-stage decision, one allows the decision to be made after several stages, much more can be achieved.

Example 5.6.2. Assume a primitive retina made of two photoreceptors. The ganglion cells can look on one receptor only (because they must see only part of the retina). Can we construct a perceptron that will detect the presence of contrast on this retina?

To state the problem differently, we built a "truth table" for the perceptron, that is, a list of all possible inputs and the desired output, as shown in Table 5.6.1. This is also known as an XOR (exclusive-OR) operation. It turns out that a perceptron *cannot* perform an XOR operation! However, a two-layered perceptron can easily do so, as shown in Figure 5.6.2. If the retina is in the dark, none of the ganglion cells fire, and therefore none of the "hidden units" fire, and the contrast detector does not fire either. If the retina is completely illuminated, both ganglion cells fire, and all the hidden units fire. Their effects on the contrast detector cancel each other, and the contrast detector does not fire. However, if one photoreceptor is illuminated and one is in the dark, a net excitation of 1 arrives at the contrast detector, and it fires.

In general, the multilayered perceptron is made of L layers. Each neuron in layer l is affected synaptically by all the neurons in layer $l - 1$, and if the added synaptic effects exceed its threshold, it will fire and affect the neurons in layer $l + 1$. The network is fed by inputs from local feature extractors (ganglion cells, in the examples), which are considered as being in layer number 0. In the last (output) layer, each neuron

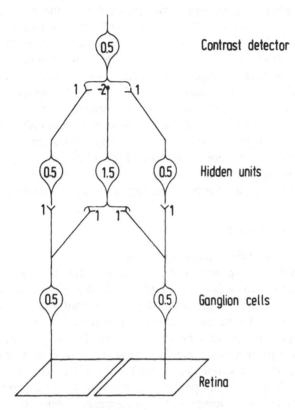

Contrast detector

Hidden units

Ganglion cells

Retina

Figure 5.6.2. Exclusive-OR operation by a perceptron with one layer of hidden units.

should become active only when some global feature is present in the input layer. The multilayered perceptrons are hierarchical structures with *no feedback*. Activity is fed only forward from one layer to the next; in each layer, one expects to find neurons that can extract some more complex or more global features, until one obtains the ultimate goal of the network at the last layer.

Two basic learning procedures have been used for multilayered perceptrons. In the first algorithm, one selects a set of input patterns as "prototypes." For each prototype, one determines the desired firing pattern for each neuron in each layer l ($l = 1, 2, \ldots , L$). Synaptic strengths are then adjusted for every neuron according to the rules described for a single perceptron.

In the other learning scheme, one measures only the error at the output (Lth) layer and uses this to correct all the synaptic efficacies of the network. This method is known as back-propagation of error [Rumelhart, Hinton, and Williams, 1986]. In this algorithm the neurons are considered as analog output devices (their output is the firing rate averaged over the synaptic integration time). An input is presented, and the output is measured. If the desired output pattern is obtained, nothing is changed. If erroneous output is obtained, one computes for every synapse (J_{ij}) its contribution to the final error (by computing the change in error that will result from a small modification of that synapse) and then modifies its strength in the direction that will reduce the total error. That is, learning is carried out by modifying the synaptic strength according to

$$\Delta J_{ij} \propto -\partial \text{error}/\partial J_{ij}$$

where ΔJ_{ij} is the "learned" change in the synaptic strength.

Learning by back-propagation of errors is attractive because there is no need to have knowledge regarding the desired properties of the hidden units. One needs only to define the desired input–output relationships of the entire network and be able to measure (calculate) the errors between the actual output and the desired output. However, in practice, learning by back-propagation typically requires many thousands of learning sessions, and quite often the network does not reach the desired performance. These practical problems are severe because there is no *convergence theorem* for multilayered perceptrons. Thus, there is no assurance that with enough trials the network will learn (even though it is easy to show that for a finite number of input patterns any classification can be carried out if enough hidden units are added).

This difficulty is overcome when using the first algorithm. In this algorithm the designer specifies for each input not only the desired output but also the desired state of each hidden unit. Therefore, each hidden unit can be treated as an isolated perceptron, and the perceptron convergence theorem assures us that if the desired state can be achieved, it will be achieved after a finite number of learning sessions. Although the convergence problem is solved by this algorithm, the need to define the desired state of every hidden unit is very demanding; see Grossman, Meir, and Domany [1988] regarding alleviation of this difficulty.

Recently, a formal algorithm for constructing multilayered perceptrons was suggested by Mezard and Nadal [1989]. This algorithm speci-

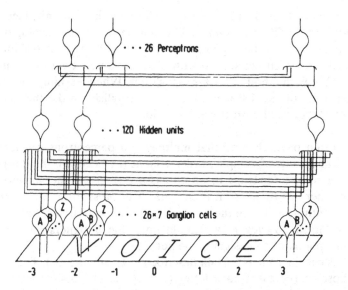

Figure 5.6.3. Structure of NET talk. Seven characters are presented on the "retina". The 26 perceptrons determine the articulation of the vocal tract required to produce the correct sound for the letter in position 0. As text glides across the "retina," the net "talks" in an intelligible form.

fies the number of layers, the number of hidden units, and the properties of the hidden units.

Example 5.6.3. Sejnowsky and Rosenberg [1987] constructed a three-layer feed-forward system that can be taught to convert written English text into phonemes. Figure 5.6.3 illustrates the structure of their network, which they call "NET talk." The purpose of the net is to convert a letter of written text into a sound. The input to the net is the letter that is to be converted, as well as the text around it. Typically, three letters before and three letters after the letter to be converted are given. The output is a layer of "command neurons" to the vocal-tract motor system that signifies the articulatory features of the tract, the voiced/unvoiced characteristic of the phoneme, the vowel height, the stress, and so forth. The input layer contains seven groups of 26 neurons each. Within each group, only one neuron is active, representing the text (letter, or punctuation) at that position. The 182 input units feed their effects onto 120 intermediate (hidden) units. Each hidden unit receives input from all 182 input units. The 120 hidden units feed their output onto the 26 output units. The neurons are of the analog output type. A network of this type has $182 \cdot 120 + 120 \cdot 26 = 25,000$ internal connections, which are started with random values. Learning is implemented by the method of back-propagation. When learning was carried out on a set of the 1,000 most common English words, the network achieved 98 percent correct pronunciation.

It has often been claimed that multilayered perceptrons can learn to solve many kinds of complex tasks without the need (on the designer's part) for prior knowledge or intuition for connecting the network. All the designer has to do, so it is claimed, is to present the input and the desired output, and then the network will learn a desired set of connections by itself. However, we should note that designers do use their preconceptions when designing the network.

Let us consider the NET talk described in Example 5.6.3. The designers chose to represent each letter (out of the 26 possible letters and punctuations) by the activity of one ganglion cell (out of the 26 cells) representing a given position in the input text. This choice was not the most efficient one. A binary representation of 26 symbols requires only five neurons. The designers of NET talk did not explain why they chose this particular mode of representing the input, but it stands to reason that a binary representation would require many more hidden units and might not converge to the desired result (I thank M. Angelo for drawing my attention to this issue).

There is great mathematical similarity between multilayered perceptrons composed of yes/no neurons and recurrent cooperative networks that are updated synchronously in parallel. Those similarities include, among other things, the immunity of multilayered perceptrons to synaptic noise, to the clipping of the synaptic strength, and to the random deletion of many synapses [Domany, Kinzel, and Meir, 1989].

5.6.3 Multilayered perceptrons and the cortex

Multilayered perceptrons work essentially in a manner similar to the prevailing neurophysiological view. According to this view, on arrival at the cortex, sensory information is subject to a hierarchy of feature extractions.

The properties of neurons in the visual cortex may serve to illustrate this view. An individual neuron in the primary visual cortex fires at a high rate when a line in a certain orientation is present in its (visual) receptive field. According to some views [Hubel and Weisel, 1962], a few neurons with simple features converge on a "complex cell" that extracts more complex features, and several complex cells converge on a "hypercomplex" cell.

More recently, a similar hierarchy for color extraction has been described [Livingstone and Hubel, 1984; Hubel and Livingstone, 1987]. There, processing starts with a simple wavelength sensitivity in the photoreceptors of the retina and ends with neurons that exhibit the perceptual property of color constancy in area V4 [Zeki, 1980]. Although there is some controversy regarding the existence of the "simple" to "complex" to "hypercomplex" hierarchy, it is generally accepted that as one moves from primary to secondary to tertiary areas, the features that make the neurons fire become more and more complex. Current models of the motor system also imply a feed-forward structure [Anderson and Zipser, 1988; Georgopoulos et al., 1989].

Although physiological recordings from sensory and motor areas seem to support the multilayered perceptron scheme, the internal anatomy of the cortex does not. A multilayered perceptron has essentially only feed-forward connections, whereas the local cortical anatomy supports a multitude of connections through which the neurons can recurrently interact with each other. Physiologists often have suggested models in which input to the cortex first affects neurons in layer IV, and from there spreads to the other layers [Eccles, 1981], and some have even provided electrophysiological evidence for that [Mitzdorf and Singer, 1978]. However, anatomy shows that thalamic afferents make contacts with all the neural processes present in layer IV, albeit with different probabilities [White, 1989]. Thus, almost all cortical neurons receive information directly from the thalamus. The local corticocortical connections (with the exception of the chandelier cell) seem to follow the same rule [Martin, 1988; White, 1989]; that is, an axon can establish synaptic contacts with all existing postsynaptic sites present in its range of distribution. Thus, the thalamic input can directly affect most of the cortical neurons, and the cortical neurons can affect each other recurrently. This is certainly not a feed-forward structure.

One way to explain this apparent inconsistency between physiology and anatomy is to assume that most of the morphological synapses are functionally ineffective [Wall, 1988]. Another way is to attribute it to the conditions (anesthesia and paralysis) under which most recordings are obtained.

The dogma stating that each neuron in the various sensory systems has a trigger feature that causes it to fire at high rates was clearly formulated by Barlow [1972], who called such neurons "cardinal cells." A cardinal cell is equivalent to the perceptron described in this section. According to the cardinal cell dogma, the relevant parameter describing the neuron's activity is its firing rate. Thus, the proper description of a neuron for modeling is that with an analog output.

However, when recording from nonprimary cortical areas, the firing rates of the responding neurons are quite low [Abeles et al., 1990]. A single neuron firing at, say, 30/s will have only a small effect on postsynaptic target neurons, whose integration time constant is less than 10 ms (particularly if one recalls that the ASG of the cortical synapse is very small). One must assume that there are always groups of neurons that elevate their firing rate, and whose outputs converge on the target neurons. With this type of arrangement, we wish to know if such a group consists of many replicas of the same cardinal cell or of a mixture of neurons with various response properties.

If a physiologist had to analyze a multilayered perceptron machine in terms of cardinal cells, he would face an easy task when dealing with the input and output layers, but an impossible task when dealing with the hidden units. For instance, in the NET talk described in Example 5.6.3, each of the input neurons can be described in terms of its receptive field (at which of the seven input characters it looks) and its trigger features (what letter it represents). Each of the output neurons can be described as a command neuron performing a certain motor output (certain articulation of the vocal tract). However, the hidden units need not be characterized by any simple "trigger feature." Instead, the input may be coded by the combination of the hidden units that fire together. If the network learned through back-propagation of errors, then there could exist many networks with identical overall input–output relations, but with very different properties of the individual hidden units.

In recent years there have been several attempts to revive the term "cell assembly" coined by Hebb [1949]. These views hold that cortical coding is not carried out by cardinal cells (or "grandmother cells," as their opponents like to call them) but by activity in cell assemblies [Hebb, 1949; Braitenberg, 1978; Palm, 1982; Gerstein, Bedenbaugh, and Aertsen, 1989]. According to this hypothesis, a feature of the external world is not represented in one neuron (or a few neurons) that specializes in detecting this feature, but rather by activity in a whole group of neurons. A given neuron may participate in several cell assemblies. Therefore, when studying the response properties of a cortical neuron, it may be pointless to search for *the adequate stimulus* that drives it.

Typically, one assumes that the neurons within a cell assembly strengthen mutual excitatory connections among themselves. In this fashion they generate an "attractor" within a recurrent neural network, as described in Section 5.5.

Note, however, that within the framework of both multilayered perceptrons and recurrent networks one can find representation of a feature by groups of neurons. Thus, it seems that in orderr to characterize how the cortex works, it is profitable to distinguish between feed-forward computations (by multilayered perceptron) and feedback computations (by recurrent cooperative networks), rather than argue over feature encoding by a single neuron versus feature encoding by activity in cell assemblies.

To summarize, Multilayered feed-forward networks are used successfully for solving various complex computational problems. The current neurophysiological prespective regarding cortical processing falls in line with their feed-forward construct. However, this construct seems to contradict the anatomical structure of the cortex, which allows for multitudes of recurrent feedbacks. The next two chapters suggest a way to reconcile this apparent conflict.

5.7 References

Abeles M., Vaadia E., and Bergman H. (1990). Firing patterns of cortical neurons and neural network models. *Network* 1:13–25.

Amit, D. J. (1989). *Modeling Brain Function.* Cambridge University Press.

Amit D. J., Gutfreund H., and Sompolinsky H. (1985a). Spin-glass models of neural networks. *Phys. Rev.* A32:1007–18.

(1985b). Storing infinite numbers of patterns in a spin-glass model of neural networks. *Phys. Rev. Lett.* 55:1530–3.

(1987a). Statistical mechanics of neural networks near saturation. *Annals of Physics* 173:30–67.

(1987b). Information storage in neural networks with low levels of activity. *Phys. Rev.* A35:2293–303.

Andersen P. and Andersson S. A. (1974). Thalamic origin of cortical rhythmic activity. In Creutzenfeldt O. (ed.), *Handbook of Electroencephalography and Clinical Neurophysiology,* Vol. 2c, pp. 90–114. Elsevier, Amsterdam.

Andersen R. A. and Zipser D. (1988). The role of the posterior parietal cortex in coordinate transformations for visual-motor integration. *Can. J. Physiol. Pharmacol.* 66:488–501.

204   5. Models of neural networks

Anninos P. A. (1972). Cyclic modes in artificial neural nets. *Kybernetik* 11:5–14.
Anninos P. A., Beek B., Csermely T. J., Harth E. M., and Pertile G. (1970). Dynamics of neural structures. *J. Theor. Biol.* 26:121–48.
Barlow, H. B. (1972). Single units and sensation: A neuron doctrine for perceptual psychology. *Perception* 1:371–94.
Beurle R. L. (1956). Properties of a mass of cells capable of regenerating pulses. *Philos. Trans. R. Soc. Lond.* [*Biol.*] 240:55–94.
Bishop G. H. and Smith J. M. (1964). The size of nerve fibers supplying the cerebral cortex. *Exp. Neurol.* 9:483–501.
Braitenberg V. (1978). Cell assemblies in the cerebral cortex. In Heim R. and Palm G. (eds.), *Lecture Notes in Biomathematics, Vol. 21: Theoretical Approaches to Complex Systems,* pp. 171–88. Springer-Verlag, Berlin.
Buhmann J. (1989). Oscillations and low firing rates in associative memory neural networks. *Phys. Rev. A* 40:4145–8.
Buhmann J. and Schulten K. (1987). Influence of noise on the function of a "physiological" neural network. *Biol. Cybern.* 56:313–27.
Burns D. B. (1958). *The Mammalian Cerebral Cortex.* Arnold, London.
Burns D. B. and Webb A. C. (1979). The correlation between discharge times of neighboring neurons in isolated cerebral cortex. *Proc. R. Soc. Lond.* [*Biol.*] 203:347–60.
Caianiello E. R. (1961). Outline of a theory of thought processes and thinking machines. *J. Theor. Biol.* 2:204–35.
Cox D. R. (1962). *Renewal Theory.* Methuen, London.
Cragg B. G. and Temperly H. N. V. (1955). Memory: The analogy with ferromagnetic hysteresis. *Brain* 78:304–416.
Crick F. (1984). The function of the thalamic reticular complex: The searchlight hypothesis. *Proc. Natl. Acad. Sci. U.S.A.* 81:4586–90.
Derrida B., Gardner E., and Zippelius A. (1987). An exactly solvable asymmetric neural netwrok model. *Europhys. Lett.* 4:167–73.
Dewdney A. K. (1984). Computer recreation: The failing of a digital eye suggests that there can be no sight without insight. *Sci. Am.* (Sept.):22–30.
Domany E., Kinzel W., and Meir R. (1989). Layered neural networks. *J. Phys. A: Math. Gen.* 22:2081–102.
Eccles J. C. (1981). The modular operation of the cerebral cortex as the material basis of mental events. *Neuroscience* 6:1839–56.
Ermentrout G. B. and Cowan J. D. (1979). A mathematical theory of visual hallucination patterns. *Biol. Cybern.* 34:137–50.
Gardner E. (1988). The space of interactions in neural network models. *J. Phys. A: Math. Gen.* 21:257–70.
Georgopoulos A. P., Larito J. T., Petrides M., Schwartz A. B., and Massey J. T. (1989). Mental rotation of the neuronal population vector. *Science* 243:234–6.
Gerstein G. L., Bedenbaugh P., and Aertsen A. M. H. J. (1989). Neuronal assemblies. *IEEE Trans. Biomed. Engin.* 36:4–14.
Gestzi T. (1990). *Physical Models of Neural Networks.* World Scientific Publishing, Singapore.

Glauber R. J. (1963). Time-dependent statistics of the Ising model. *J. Math. Phys.* 4:294–307.

Griffith J. S. (1963). On the stability of brain-like structures. *Biophys. J.* 3:299–308.

— (1971). *Mathematical Neurobiology: An Introduction to the Mathematics of the Nervous System* Academic Press, New York.

Grossman T., Meir R., and Domany E. (1988). Learning by choice of internal representation. *Complex Systems* 2:287.

Harth E. M., Csermly T. J., Beek B., and Lindsay R. D. (1970). Brain function and neural dynamics. *J. Theor. Biol.* 26:93–120.

Haug H., Kolln M., and Rost A. (1976). The postnatal development of myelinated nerve fibers in the visual cortex of the cat. *Cell Tissue Res.* 167:265–88.

Hebb D. O. (1949). *The Organization of Behavior.* Wiley, New York.

Hopfield J. J. (1982). Neural networks and physical systems with emergent collective computational abilities. *Proc. Natl. Acad. Sci. U.S.A.* 79:2554–8.

— (1984). Neurons with gradual response have collective properties like those of two-state neurons. *Proc. Natl. Acad. Sci. U.S.A.* 81:3088–92.

Hopfield J. J. and Tank D. W. (1986). Computing with neural circuits: A model. *Science* 233:625–33.

Hubel D. H. and Livingstone M. S. (1987). Segregation of form, color, and stereopsis in primate area 18. *J. Neurosci.* 7:3378–415.

Hubel D. H. and Wiesel T. N. (1962). Receptive fields, binocular interactions and functional architecture in the cat's visual cortex. *J. Physiol. (Lond.)* 160:106–54.

John E. R. (1972). Switchboard versus statistical theories of learning and memory. *Science* 177:850–64.

Katznelson R. D. (1981). Normal modes of the brain: Neuroanatomical basis and physiological theoretical model. In Nunez P. L. (ed.), *Electric Fields of the Brain: The Neurophysics of EEG,* pp. 401–42. Oxford University Press.

Kohonen T. (1984). *Self-organization and Associative Memory.* Springer-Verlag, Berlin.

Krone G., Mallot H., Palm G., and Schuz A. (1986). Spatiotemporal receptive fields: A dynamical model derived from cortical architectonics. *Proc. R. Soc. Lond. [Biol.]* 226:421–44.

Lashley K. S. (1929). *Brain Mechanisms and Intelligence.* University of Chicago Press.

Little W. A. (1974). The existence of persistent states in the brain. *Math. Biosci.* 19:101–20.

Livingstone M. S. and Hubel D. H. (1984). Specificity of intrinsic connections in primate primary visual cortex. *J. Neurosci.* 4:2830–5.

McCulloch W. S. and Pitts W. (1943). A logical calculus of ideas immanent in nervous activity. *Bull. Math. Biophys.* 5:115–33.

MacGregor R. J. (1987). *Neural and Brain Modeling.* Academic Press, San Diego.

Martin K. A. C. (1988). From single cells to simple circuits in the cerebral cortex. *Q. J. Exp. Physiol.* 73:637–702.

Mezard M. and Nadal J. (1989). Learning in feedforward networks: The tiling algorithm. *J. Phys. A: Math. Gen.* 22:2191–203.

Minsky M. and Papert S. (1969). *Perceptrons.* MIT Press, Cambridge, Mass.

Mitzdorf U. and Singer W. (1978). Prominent excitatory pathways in the cat visual cortex (A17 and A18): A current source density analysis of electrically evoked potentials. *Exp. Brain Res.* 33:371–94.

Nadal J. P., Toulouse G., Changeux J. P., and Dehaene S. (1986). Networks of formal neurons and memory palimpsests. *Europhys. Lett.* 1:535–42.

Nelken I. (1985). Analysis of the activity of a single neuron in a random net. M.Sc. thesis, The Hebrew University, Jerusalem.

(1988). Analysis of the activity of single neurons in stochastic settings. *Biol. Cybern.* 59:201–15.

Palm G. (1982). *Studies of Brain Function, Vol. VII: Neural Assemblies.* Springer-Verlag, Berlin.

(1986a). Associative networks and cell assemblies. In Palm G. and Aertsen A. (eds.), *Brain Theory*, pp. 211–28. Springer-Verlag, Berlin.

(1986b). Warren McCulloch and Walter Pitts: A logical calculus of ideas immanent in nervous activity. In Palm G. and Aertsen A. (eds.), *Brain Theory.* pp. 229–30. Springer-Verlag, Berlin.

Pantiliat S. (1985). Theoretical model of the nervous system. Ph.D. thesis, The Hebrew University, Jerusalem.

Peretto P. (1989). *The Modeling of Neural Networks.* Editions de Physique, Les Ulis.

Pitts W. and McCulloch W. S. (1947). How do we know universals: The perception of auditory and visual forms. *Bull. Math. Biophys.* 9:127–47.

Rosenblatt F. (1962). *Principles of Neurodynamics.* Spartan, New York.

Rubin N. and Sompolinsky H. (1989). Neural networks with low local firing rates. *Europhys. Lett.* 10:465–70.

Rumelhart D. E., Hinton G. E., and Williams R. J. (1986). Learning representations by back-propagating errors. *Nature (Lond.)* 323:533–6.

Rumelhart D. E. & McClelland J. L. (eds.) (1986). *Parallel Distributed Processing: Exploration in the Microstructure of Cognitions*, 2 vols. MIT Press, Cambridge, Mass.

Sejnowsky T. J. and Rosenberg C. R. (1987). Parallel networks that learn to pronounce English text. *Complex Systems* 1:145–68.

Shinomoto S. (1987). A cognitive and associative memory. *Biol. Cybern.* 57:197–206.

Sholl D. A. (1956). *The Organization of the Cerebral Cortex.* Methuen, London.

Sillito A. M., Kemp J. A., Milson J. A., and Bernardi N. (1980). A re-evaluation of the mechanisms underlying simple cell orientation selectivity. *Brain Res.* 194:517–20.

Sompolinsky H., Crisanti A., and Sommers H. J. (1988). Chaos in random neural networks. *Phys. Rev. Lett.* 61:259–62.

Takizawa N. and Oonuki M. (1980). Analysis of synapse density in the cerebral cortex. *J. Theor. Biol.* 82:573–90.

Tomasch J. (1954). Distribution and number of fibers in the human corpus callosum. *Anat. Rec.* 119:119–35.

Toulouse G. (1989). Perspectives on neural network models and their relevance to neurobiology. *J. Phys. A: Math. Gen.* 22:1959–68.

Treves A. and Amit D. J. (1989). Low firing rates: An effective Hamiltonian for excitatory neurons. *J. Phys. A: Math. Gen.* 22:2205–26.

Wall P. D. (1988). Recruitment of ineffective synapses after injury. *Adv. Neurosci.* 47:387–400.

White E. L. (1989). *Cortical Circuits.* Birkhauser, Boston.

Wilson H. R. and Cowan J. D. (1973). A mathematical theory of the functional dynamics of cortical and thalamic nervous tissue. *Kybernetik* 13:55–80.

Wooldridge D. E. (1979). *Sensory Processing in the Brain: An Exercise in Neuro-connective Modeling.* Wiley, New York.

Zeki S. (1980). The representation of colours in the cerebral cortex. *Nature (Lond.)* 284:412–18.

6

Transmission through chains of neurons

6.1 The necessity of diverging/converging connections

The execution of brain processes often requires hundreds of milliseconds. Even a simple reaction-time exercise (in which the subject is required to press a button each time a sound is heard) has a central delay of about 100 ms. As the task becomes more complex (e.g., by requesting the subject to respond only to one of several possible sounds), the central delay becomes longer. The long central delay is accounted for by assuming that information processing is done through neuronal activity that must traverse through many stations in tandem. This type of processing often is visualized as being carried through a chain of serially connected neurons, such as that shown in Figure 6.1.1A. However, this arrangement is faulty, because if one of the neurons in the chain is damaged, the entire chain (composed of n cells) will become inoperative. The cortex is subject to a process by which its neuronal population is continually thinned out. Comparisons of neuronal densities in the brains of people who died at different ages (from causes not associated with brain damage) indicate that about a third of the cortical cells die between the ages of twenty and eighty years [Gerald, Tomlinson, and Gibson, 1980]. Adults can no longer generate new neurons, and therefore those neurons that die are never replaced.

The neuronal fallout proceeds at a roughly steady rate throughout adulthood (although it is accelerated when the circulation of blood in the brain is impaired). The rate of neuronal fallout is not homogeneous throughout all the cortical regions, but most of the cortical regions are affected by it.

Let us assume that every year about 0.5 percent of the cortical cells die at random. This means that within ten years we can expect five cells in a chain of 100 cells to be missing. Although five cells are expected to die, it is possible that by a lucky chance none will. We can

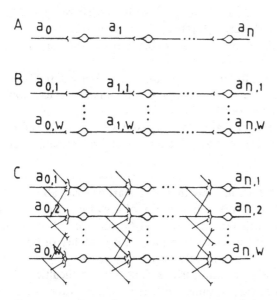

Figure 6.1.1. Connections between neurons in a chain. A: Simple chain with *n* synapses. B: Parallel chains. C: Diverging/converging chain.

compute the probability of having zero deaths when five deaths are expected considering the following: We face a large number of candidates (100), each of which has a low (0.05) probability of dying. In these conditions, the probability of finding *n* deaths is given by the Poisson formula

$$\mathrm{pr}\{n;\, x\} = e^{-x}x^n/n! \tag{6.1.1}$$

This formula states the probability of observing, by chance, *n* occurrences of an event when we expect to observe *x* occurrences. In our case, we are interested in knowing the probability of finding no cell death ($n = 0$) when the expected number of deaths is five ($x = 5$). This probability is

$$\mathrm{pr}\{0;\, 5\} = e^{-5} = 0.007$$

This means that almost all the chains that are 100 (or more) synapses long will become inoperative. A simplistic way to overcome this difficulty would be to think of many serial chains, such as in Figure 6.1.1A, operating in parallel (Figure 6.1.1B), but the following exercise will show that not to be a practical solution.

Exercise 6.1.1. A brain process is carried out by w parallel chains, as shown in Figure 6.1.1B. Let us assume that each chain is 100 cells long, that cells fall out at a rate of 0.5 percent per year, and that this fallout is completely random.

1. If we wish to assure ourselves (with a probability of 0.99) that after ten years at least one of the w chains will remain intact, how many parallel chains do we need?

2. How many parallel chains do we need if we want to be certain (with a probability of 0.99) that at least one chain will remain intact after twenty years?

Solving Exercise 6.1.1 will show that if we start out with about 600 serial chains (or more), we have a probability of 0.99 that after ten years at least one of them will remain intact. However, we will need at least 60,000 parallel chains to guarantee (with a probability of 0.99) that at least one chain will remain intact after twenty years.

Serial chains also present serious functional difficulties. In these chains, each of the synapses must be very strong in order to ensure that spikes that start off at one end of the chain will be transmitted all the way through to the other end. This issue was discussed in Chapter 3, where we concluded that such a form of transmission is highly unlikely for the cortex. In addition, we showed there (Exercise 3.5.1) that the dispersion of the excitation after so many synapses (even if they were strong enough) would be enormous.

It seems that one can overcome the previously mentioned problems by allowing the axon of each neuron in each station to branch out and make excitatory synapses with several of the cells in the next station along the chain. In such an arrangement, every neuron in a station excites several neurons in the next one, and every neuron in the next station receives excitation from several neurons of the previous one.

We have reached a very important conclusion: In the cortex, reliable transmission of activity is possible only between populations of cells connected by multiple diverging and converging connections.

An example of such a chain is shown in Figure 6.1.1C, where the chain consists of nodes, and each of the cells in one node branches off and contacts three of the cells in the next node, and each of the cells in the next node receives contacts from three cells of the preceding one.

Griffith [1963] has suggested that such chains, in which each cell in one node excites all the other cells in the next node, provide a good way of transmitting excitation through a network of otherwise randomly connected excitatory cells, without failure and without activating most of the other cells in the net. He has called this type of chain a "complete transmission line."

Let us now examine how many diverging connections each cell in the exciting node should give in order to ensure transmission along the chain. Let us assume that each cell branches off and makes excitatory contacts with m cells in the next node, and let us denote the strength (ASG) of each connection by a. When such a cell fires, it will generate, on the average, ma spikes in the receiving node. Thus, we should have

$$ma = 1 \qquad (6.1.2)$$

in order to ensure unattenuated transmission along the chain.

This relation gives us the range of branching that we should look for in the cortex. In Chapter 3 we found that the ASG of cortical synapses was in the range of 0.003 to 0.2. Thus, for $a = 0.003$, we require $m = 330$ and for $a = 0.2$, we require $m = 5$. Therefore, the range of divergence required in order to preserve activity along a chain is 5 to 330.

Cortical cells give synapses to thousands of their neighbors. Therefore, the amount of divergence required to maintain transmission obviously is present in the cortex. However, equation (6.1.2) deals only with the degree of divergence required to maintain the average level of activity along the chain. It does not specify any degree of convergence. Indeed, we can imagine a situation in which the first node is 10 cells wide, and it gives off diverging connections to a node that is 100 cells wide, which in turn gives off diverging connections to a node 1,000 cells wide. If the connections meet the requirement $ma = 1$, then we may face a situation in which activation of all the 10 cells in the first node will excite only 10 (out of 100) cells in the second node, which will excite only 10 (out of 1,000) cells in the third node. In this situation, if we excite all the 10 cells of the first node repeatedly, we shall obtain repeated excitation of 10 cells in the following nodes, but these 10 cells will be different for almost every repeated activation. In such cases we cannot speak of reproducible transmission along a chain of neurons. Therefore, we limit our consideration to chains in which the width of each node (w) is constant. Thus, the average convergence of excitatory connections on a cell from the receiving node must be equal to the average divergence of the branches of a cell in the sending node.

We call such chains of neurons diverging/converging chains. Each station in such a chain will be called a node. The number of cells in each node is the width of the chain (w), and the degree of divergence and convergence will be called the multiplicity (m) of connection between adjacent nodes. (In Figure 6.1.1C the multiplicity of connections is three.) The case in which $w = m$ (i.e., every cell in one node excites every cell in the next node) is Griffith's complete chain.

Our calculations [based on equation (6.1.2) and the discussion in Section 3.3.1] suggest that a cortical diverging/converging chain can function if the multiplicity of the connections is between 5 and 330.

Configurations such as that in Figure 6.1.1C may suggest that each node in the chain is placed in a separate anatomical location, but that probably is not true in the cortex. The arrangement of the cells in Figure 6.1.1C represents their time of activation. The reader should imagine taking these cells and distributing them at random among tens of thousands of other neurons, without disrupting their connections.

The entire concept of transmitting activity through a chain of neurons may seem trivial. If the activity is presented in the first group of w neurons, why bother to transmit it through 100 links just to get it into another group of w neurons? Again, configurations such as Figure 6.1.1C are misleading. Imagine that a certain piece of information arrives at a cortical region and is processed there for a while, and then the results are emitted as output. If during the time that the process occurs the total amount of activity in the region remains constant, one can divide the cells that participate in it into groups. Those likely to be active in the first stage (of 1 ms or so) are in group 0; those that are likely to be active in the second stage are in group 1; and so forth. These groups are then depicted in a manner similar to Figure 6.1.1C. Note that if the processing includes multiple feedbacks of excitation, then a given neuron will appear in more than one group along the "chain." Consequently, Figure 6.1.1C should not be regarded as a transmission line but rather as a time diagram of the evolvement of activity within a network.

To summarize, we have argued that transmission of information in the cortex must take place through diverging/converging connections between populations of neuron. We have estimated that the degree of convergence/divergence (the multiplicity) of the connections should be in the range of 5 to 330. In the next section we examine the probability that such connections exist in the cortex.

6.2 Existence of diverging/converging connections

In this section we examine the plausibility of having diverging/converging connections between two populations of neurons, as described in the preceding section. The procedure for appraising this plausibility is as follows: Let us assume that we have found one set (the sending node) of neurons, and its width is w (it is composed of w neurons). We shall compute the probability of finding, in the vicinity of these neurons, at least w other neurons (the receiving node) that are linked to the original node thorugh diverging/converging connections with a multiplicity of at least m. If the probability of finding such a receiving node is high (close to 1), then it is reasonable to assume that diverging/converging chains of width w and multiplicity m exist in abudance. If, on the other hand, the probability of finding the receiving node is very low (close to zero), then we conclude that this type of link is not likely to be formed by chance. That is, either they do not exist at all or there are special growth rules that favor the formation of diverging/converging connections.

In order to illustrate how we can compute such probabilities, we adopt the simplest mode of local connectivity, which was discussed in Chapter 2. Let us assume that our neurons are distributed in a small cortical volume (0.5 mm^3) that contains 20,000 neurons, each of which forms excitatory contacts with 5,000 other neurons randomly. It must be clear that this is an oversimplification and that in the future more accurate probabilities should be calculated using the more realistic connectivity patterns described in Chapter 2.

6.2.1 Complete chains of Griffith

Complete chains are those for which $m = w$. That is, every cell in the sending node sends excitation to all the cells in the receiving node, and every cell in the receiving node receives excitatory input from all the cells in the sending node. Complete chains of widths 1, 2, and 3 are shown in Figure 6.2.1. In the trivial case $w = m = 1$ (Figure 6.2.1A), every node is composed of only one cell, which excites the cell in the next node. When $w = m = 2$ (Figure 6.2.1B), every node is composed of two cells, each of which excites both cells in the next node. Similarly, for $w = m = 3$ (Figure 6.2.1C), each of the three cells in one node excites all of the three cells in the next node.

What is the probability of such connections being formed by chance in a network composed of 20,000 cells in which every cell has a probability of 0.25 to excite any other cell? It is obvious that there is no problem in obtaining such connections for $w = m = 1$, because for every cell we choose, there will be at least one cell that will be excited by it and therefore may serve as the next node in the chain.

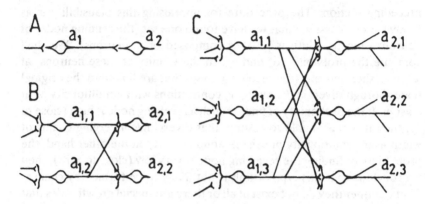

Figure 6.2.1. Complete diverging/converging connections. A: Chain width of 1. B: Chain width of 2. C: Chain width of 3.

What about $w = m = 2$? Suppose we select $a_{1,1}$ and $a_{1,2}$ in Figure 6.2.1B as the sending node. What is the probability of finding at least two cells that will be excited by both of them and can be used as the receiving node? Axon $a_{1,1}$ branches off and excites 5,000 other neurons in the net. Every one of these has a probability of 0.25 of being excited by cell $a_{1,2}$ as well, so that we can expect to find 1,250 cells that are excited by both $a_{1,1}$ and $a_{1,2}$. Although 1,250 is a large number compared with 2, it is still possible that by chance no cell will be excited by both cells of the sending node, or that only one such cell will be. What is the probability in these cases? The exact probability of finding only zero or one such cell is given by the binomial distribution, but because there are many candidates (20,000) and the probability for any one of them is small (particularly as the degree of divergence/convergence becomes high), we can again use the Poisson formula:

$$\mathrm{pr}\{n; x\} = e^{-x}x^n/n!$$

which for the present case will yield the following probabilities for finding no cell or finding only one cell:

$$\mathrm{pr}\{0; 1{,}250\} = e^{-1{,}250} \approx 10^{-543}$$
$$\mathrm{pr}\{1; 1{,}250\} = 1{,}250e^{-1{,}250} \approx 10^{-540}$$

the probability that there will be at least two cells that are excited by both cells from the sending node is

$$1 - \mathrm{pr}\{0; 1{,}250\} - \mathrm{pr}\{1; 1{,}250\} \approx 1$$

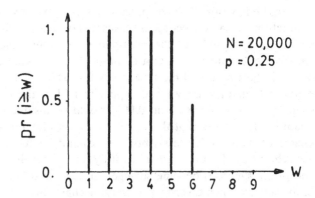

Figure 6.2.2. Probability of finding complete diverging/converging connections as a function of the number of cells (w) in each node.

Thus, we can easily find two cells that receive a complete set of connections from the first node. For any pair of cells that we select to serve as the second node, there will again be no difficulty in finding another pair that receive a complete set of connections from that (second) pair, and so on. Therefore, there is no difficulty in finding many complete chains in the network whose width is 2.

In general, to calculate the probability of finding a complete link whose width is w, we proceed in the following way:

1. The probability that all of the w cells in the first node will excite any given cell is

$$\text{pr}\{w\} = 0.25^w$$

2. The expected number of cells that will receive excitation from all the cells in the first node is

$$x = N \, \text{pr}\{w\} = 20{,}000 \cdot 0.25^w$$

3. The probability of finding w or more such cells is

$$\text{pr}\{I \geq w\} = 1 - \text{pr}\{i < w\} = 1 - \sum_{i=0}^{w-1} \text{pr}\{i; x\} \tag{6.2.1}$$

Figure 6.2.2 shows a graphic representation of this probability as a function of the link width (w). We see here that there is no problem in finding complete diverging/converging links of widths up to 5, but there are extremely low chances of finding such chains whose widths are greater than 6.

Can such chains be functional? In the preceding section we saw that in a first approximation we can expect secure transmission along a chain if the strength of the synapses and the multiplicity of connections are related by $m\mathrm{ASG} \geq 1$. This means that a complete, 5-cell-wide chain can be operative if the strength of the synapses is at least 0.2. This value is near the upper limit of what we can expect to find in the cortex. Regular synapses, whose gain is about 0.003, would be insufficient to make the chain work. We see that the cortical network may be full of complete chains of width 5 (or less), but they cannot be functional. Some of these chains can become functional if by some "learning" process the appropriate synapses become stronger. In Section 6.3 we deal further with the way in which diverging/converging chains may operate. Here we proceed to analyze the probability of the existence of incomplete chains.

6.2.2 Incomplete chains

There is no particular reason for treating only complete chains (i.e., chains in which each cell in one node excites every cell in the next node). Spikes can also securely traverse chains in which each of the cells in the sending node contacts only some of the cells in the receiving node ($m < w$). Figure 6.2.3 illustrates such a chain, in which the width is 5, but the multiplicity of connections is only 3. By stating that the chain is of width 5 and multiplicity 3, we mean that each node has five cells, each of which is exciting at least three cells in the next node, and that in the next node each cell receives excitatory connections from at least three of the cells in the previous node.

Exercise 6.2.1. To ensure the connectivity of Figure 6.2.3, would it be sufficient to require that each cell in the sending node contact (at least) three cells in the receiving node?

In order to estimate the likelihood of finding chains of width w and multiplicity m, we calculate the probability of finding at least w cells, each of which receives at least m connections from a given set of w cells. To fix our ideas, let us calculate the probability of finding connections such as in Figure 6.2.3.

Example 6.2.1. What is the probability of finding a diverging/converging link with $w = 5$ and $m = 3$?

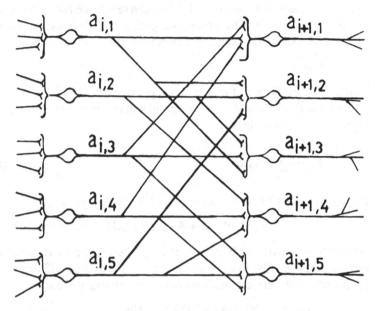

Figure 6.2.3. A patial diverging/converging link ($w = 5$, $m = 3$).

Let us choose any five cells and designate them as the sending node. Now let us pick one random cell. What is the probability that this cell will receive excitation from at least three of the five cells?

The probability that the receiving cell will be excited by one particular cell from the sending node is

$$p = 0.25$$

Its probability of not being excited by one particular cell from the sending node is

$$q = 1 - p$$

The probability that the receiving cell will be excited by only one particular cell from the sending node is

$$pq^{w-1}$$

In similar fashion we can find that the probability that the receiving cell will be excited by exactly i particular cells from the sending node is

$$p^i q^{w-1}$$

There are exactly $\binom{w}{i}$ ways in which one can select i cells out of the w cells of the sending node. Therefore, the probability that the receiving cell will receive excitation from any i cells from the sending node is

$$\binom{w}{i} p^i q^{w-i}$$

which is the ith term in the expansion of $(p + q)^w$.

The probability that the receiving cell will be excited by at least m cells from the sending node is

$$P_m = \sum_{i=m}^{w} \binom{w}{i} p^i q^{w-i} \tag{6.2.2}$$

In our example (for $w = 5$, $m = 3$, and $p = 0.25$) we get

$$P_3 = 10 \cdot 0.25^3 \cdot 0.75^2 + 5 \cdot 0.25^4 \cdot 0.75 + 0.25^5 = 0.1035$$

All in all, we have 20,000 cells that might be excited by the cells of the sending node, so that the expected number of cells that will receive three (or more) excitatory connections from the sending node is

$$x = NP_3 = 20,000 \cdot 0.1035 = 2,070$$

There should be no problem in finding at least five cells that receive the required amount of excitation.

Equation (6.2.2) is based on the binomial probability density function. When w becomes large [so that $wp \gg (wpq)^{1/2}$], the function has the form of a normal Gaussian probability density function, with an average wp and a variance of wpq.

In Table 6.2.1, x (the expected number of cells) is presented as a function of w (the width of the sending node) and m (the multiplicity of the connections). If we look at one column in the table, we are observing a node with a constant width. As an example, let us look at a column of a node that is fifteen cells wide. Following this column from the top down, we see that if all we want is to find cells that are connected to five (or more) cells of the sending node, we can expect to find 6,270 such cells; but if we want the receiving cells to receive convergent excitation from at least six cells of the sending node, we can expect to find 2,967 such cells. If we increase the degree of convergence to 7, the expected number drops to 1,132, and so on. As our requirement reaches a multiplicity of 10, we can expect to find about sixteen such cells in the cortex, and thereafter the expected number drops very quickly to almost zero.

Table 6.2.1. *Expected numbers of cells that will receive converging excitation of multiplicity m (rows) as a function of the width w (columns) of the node that sends those connections*[a]

Multiplicity of connections (m)	Width of the giving node (w)								
	5	6	7	8	10	15	20	30	50
5	19.5	92.8	257.6	546.0	978.5	6,270	11,703	18,043	19,958
6	–	4.9	26.9	84.5	394.6	2,967	7,657	15,948	19,859
7	–	–	1.2	7.6	70.1	1,132	4,284	13,039	19,612
8	–	–	–	.3	8.3	346	2,036	9,714	18,936
9	–	–	–	–	.6	84	819	6,528	18,168
10	–	–	–	–	.0	15.9	277	3,932	16,726
11	–	–	–	–	–	2.3	78.8	2,114	14,755
12	–	–	–	–	–	.2	18.7	1,013	12,367
13	–	–	–	–	–	.0	3.7	432	9,780
14	–	–	–	–	–	.0	.6	164	7,259
15	–	–	–	–	–	.0	.1	55.0	5,038
16	–	–	–	–	–	–	.0	16.4	2,847
17	–	–	–	–	–	–	.0	4.3	1,966
20	–	–	–	–	–	–	.0	.0	279
25	–	–	–	–	–	–	–	.0	2.5
30	–	–	–	–	–	–	–	.0	.0

[a]The table gives sample values for a network of 20,000 cells with a probability of contact of 0.25. For $w = 50$, the calculations are based on the normal approximation to the binomial curve.

The table does not show any figures for a multiplicity above 15 in this column, because nodes that are fifteen cells wide cannot be linked with diverging/converging connections whose multiplicity is more than 15. The sharp drop in the expected number of cells around the row $m = 10$ suggests that it is not likely that we will find diverging/converging chains of width 15 with a multiplicity of connection above 10.

Exercise 6.2.2. We select a set of w cells that send excitatory connections to the other cells in a network of 20,000 cells. The probability of one cell contacting another is 0.25. We are looking for cells that receive converging connections from at least ten of the w cells in this set. Draw a chart of the expected number of these cells as a function of w.

If we look at one row in Table 6.2.1, we see that for a given multiplicity of connections the expected number of cells is very low and rises suddenly when w becomes large enough. The place at which $x \approx w$ is the

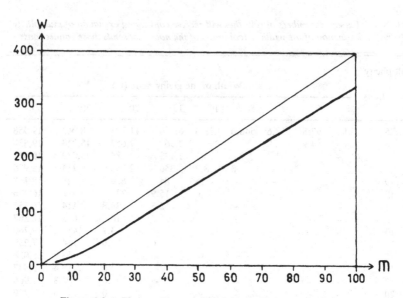

Figure 6.2.4. The smallest number (w) of cells in the sending node that will ensure a diverging/converging link of multiplicity m (heavy line). Computed for a network of 20,000 cells, with a probability of 0.25 for contact between any two cells. The thin line gives the relation $m = 0.25w$.

region where diverging/converging chains begin to be feasible. Figure 6.2.4 shows the width of the feasible chain for each multiplicity of connections. We see that even for weak synaptic contacts it is possible to form chains in which enough diverging/converging connections will exist to ensure transmission.

Exercise 6.2.3. Draw a chart of the probability of finding a diverging/converging link in which $m/w = 0.5$ (i.e., each cell in one node is connected with at least half the cells in the other node), as a function of w.

The diverging/converging link is a special case of a connectivity that might be formed by chance within a net of randomly connected vertices. The abrupt change from probable to improbable connections (Figure 6.2.2, Table 6.2.1, and Exercise 6.2.3) is a general property of such random nets [Palmer, 1985].

In the next section we discuss in detail the form of transmission through such chains. However, before proceeding to discussion of the functional aspects of the diverging/converging chain, it is appropriate to remind the reader that the probabilities in this chapter were calculated for a limited model in which every cell in the net had the same probability of contacting all the other cells. In the cortex, the connectivity is more complicated, even if we limit our discussion to pyramidal cells, and even if we assume that the axon of one pyramidal cell has the same probability of forming synapses with all the dendrites of other pyramidal cells it encounters. According to some views, the major local projections of superficial pyramidal cells are in the deep layers, whereas the major local projections of deep pyramidal cells are in the superficial layers. If this is true, then successive nodes are likely to alternate between superficial and deep layers. According to other views, the major local projection of a pyramidal cell is below its basal dendrites. If that is true, then successive nodes will tend to be below each other. Even if one assumes complete local projection (i.e., the local axonal range of a pyramidal cell overlaps with its own basal dendritic domain), the probability of contact between two pyramidal cells will vary according to the distance between them. For all these reasons, the probabilities of the existence of a diverging/converging link given earlier are only crude approximations.

We have no estimates for how the exact anatomical considerations affect the probabilities of contact and of the existence of the diverging/converging chains. But as long as the degree of divergence is considerably smaller than the total population to which one cell is connected (i.e., $m >> 5{,}000$), then two of the properties described earlier probably will remain valid: (1) For any multiplicity (m) of divergence/convergence, it will be possible to find a chain width (w) large enough to ensure its existence. (2) For any given multiplicity of connections, as the node width (w) is increased, the probability of existence will abruptly change from low to high values.

6.3 Transmission through diverging/converging chains

In this section we qualitatively examine the forms in which activity may be transmitted through diverging/converging chains. First we examine the question of stability of transmission through the chain.

6.3.1 Stability of transmission

Let us assume that we have a diverging/converging chain with many nodes and that each node is composed of w cells (the width of the chain is w). Each cell in one node (the sending node) gives synapses to m of the cells in the next node (the receiving node), and each cell in the receiving node receives synapses from m cells in the sending node (the link between two nodes is of multiplicity m). Let us further assume that the strength of these connections is

$$ASG = 1/m$$

We now examine whether or not activity can be transmitted through this chain in a stable way.

As a starting point, let us examine the simple case in which only one cell fires in the first node of this chain. We can expect that this one spike will elicit mASG spikes in the next node. Because we choose the appropriate ASG, this one spike in the first node will elicit an average of one spike in the second node, which will elicit an average of one spike in the third node, and so on. Thus, on the average, the transmission of one spike can be carried throughout the chain.

However, in real life the actual number of spikes elicited will fluctuate around the expected value. If by chance none of the excited cells reaches threshold, so that in the next node we obtain no spike (instead of the expected one spike), transmission through the rest of the chain is stopped. If, on the other hand, we obtain two spikes in the next node, then in the third node several cells will accept two partially synchronized EPSPs. In Section 4.5 we saw that for small synaptic potentials, the gain of two (partially) superposed EPSPs was stronger than the sum of their individual gains. Therefore, we can expect that the two spikes from the second node will elicit more than two spikes in the third node. If they succeed in eliciting three spikes in the third node, there will be three overlapping EPSPs on some of the cells in the fourthnode, whose gain will be even stronger, and so on.

This simple argument shows that diverging/converging chains will not transmit partially saturated activity in a stable manner. That is, for every synaptic strength (around $1/m$) there will be some marginal level of activity below which activity, as it passes from one node to the next, will diminish until it is extinguished completely. Above this level, activity will build up until it reaches saturation (i.e., all the cells in each node will fire action potentials).

This behavior will be examined again in greater detail in Chapter 7. We now proceed to examine transmission through diverging/converging chains as it evolves in the course of time.

6.3.2 Gradual transmission between nodes

Let us examine a case in which the first node in a diverging/converging chain receives excitatory input (say from a sensory source) that raises its firing rate for some T ms. After one synaptic delay (t), this activity will begin to excite the second node, and it will continue for as long as the first node is excited. The third node will be activated after two synaptic delays, and so on. If we wait for n synaptic delays, and the total delay is shorter than the duration of activation of the first group (nt < T), we get a situation in which n nodes of the chain are simultaneously active.

If the anatomy of the chain is such that every node is in a different cortical region, then this new situation (where n nodes are coactivated) poses no problem. However, the connectivity arguments presented earlier in this chapter suggested that diverging/converging connections are likely to be formed within a small local net of cortical neurons. In this case, the coactivation of many nodes in a chain poses two problems: What is the effect of inhibitory feedback on the activity? What is the reproducibility of propagation when the chain is activated repeatedly?

In Chapter 5 we saw that in the cortex we must have some local inhibitory network to prevent the cells from exciting each other to the degree that all the cells become active. When more and more nodes in the chain are recruited, this inhibitory feedback must be strongly activated. It is not clear what the effect of such inhibition will be on the activity and on the transmission through the chain.

When several nodes are coactivated, we can consider them as a new, wider node. As shown in the preceding section, wider nodes are connected to many other cells by diverging/converging connections of a higher multiplicity. This means that the coactivated nodes will excite many other cells in the vicinity that were not part of the original chain. Because the number of coactivated nodes will depend on the duration T of the original stimulus, we face a situation in which a given type of stimulus can spread into different populations in an uncontrollable manner. This type of behavior defies the very reasons for postulating the existence of diverging/converging chains.

This paradox can be summarized in the following way: In Section 6.1 we found that transmission through the cortex must be carried out between groups of cells that are connected by diverging/converging links. In Section 6.2 we found that such connections indeed exist among groups of cells in a local cortical network. However, in this section, we just found out that activity that is carried by asynchronous spatial and temporal summation of excitation cannot be transmitted through a local cortical network in an orderly, reproducible way.

We must then ask how activity can be transmitted through the cortex. There are two simple ways to resolve this problem: an anatomical solution and a functional solution. Anatomically, it is possible that the random connectivity assumed in Section 6.2 is completely wrong. If the anatomical connectivity is such that successive nodes are in different locations (layers, columns, deep nuclei), then there is no problem in transmitting asynchronous activity along the chain. Physiologically, it is possible to transmit activity along a chain whose members are dispersed in the same anatomical region, if transmission is carried by synchronous activation of all the cells in a node (and not by asynchronous integration of excitation); this alternative is described in the next section.

6.3.3 Synchronous transmission from node to node

Let us suppose that many of the cells in one node of the chain fire synchronously (within 1 ms or so). This is likely to occur when the activity in the node reaches saturation, but it can occur earlier. We have seen that the gain of several synchronous EPSPs is much larger than the sum of their ASGs. Therefore, the synchronous volley of excitation arriving at the next node is likely to synchronously excite a larger fraction of the receiving node. These cells will synchronously excite the cells in the third node, where even more cells will fire synchronously, and so forth. Soon we shall reach a situation in which all the cells in one node excite the next one synchronously, and therefore all the cells in that (the receiving) node fire synchronously, and so on.

We call this state, in which all the cells in one node fire synchronously, and after one synaptic delay all the cells in the next node fire synchronously, the *synfire* mode of transmission between nodes.

When a diverging/converging chain uses the synfire mode of operation, the problems described in Section 6.3.2 do not arise. Because every node may fire essentially only once, there is not much increase in the total activity in the region. Therefore, there is no massive inhibition to be fed back to the region.

Furthermore, because each node may fire only once before it transmits activity to the next node, it is possible to avoid the situation in which many nodes are concomitantly activated. Therefore, it is possible to avoid the activation of stray diverging/converging chains (with weak synapses), which if not avoided could result in an unpredictable evolvement of activity. Because synchronous transmission needs a lower synaptic gain than asynchronous transmission, the local inhibitory feedback can automatically prevent the chain from becoming active in its asynchronous mode.

It seems that the synfire mode of transmission through a diverging/converging chain is a natural way by which information can be processed locally in the cortex. We examine this mode in detail in Chapter 7.

To summarize, we have seen that in the cortex, activity can be transmitted only between sets of cells with multiple diverging/converging connections. We have shown that such connectivity exists and have described some of the constraints on the size of the sets and the multiplicity of the connections among them. We have suggested that a chain of such connections can operate repeatedly if the activity along such a chain of connections is transmitted by synchronous volleys of spikes.

6.4. Appendix

This appendix attempts to define more precisely the sets of neurons discussed in the preceding sections. It is intended mostly to serve as an exercise in defining sets and describing the relations between them. Here we deal with a local network of excitatory cells. The network is assumed to contain a large, but finite, number of cells. The set of all the cells in this network will be denoted by Ω. Let us choose a subset (W) of neurons containing w neurons. For any given W, we can define two associated sets of cells: the set of all cells in Ω that send synaptic contacts to neurons in W, and the set of all cells in Ω that receive synaptic contacts from the cells in W.

Definition 6.4.1: The sending set of multiplicity m. The set of all the cells in Ω each of which sends synapses to at least m cells in W is called "the sending set of multiplicity m." This set is denoted by S_m. Any subset of S_m is called "a sending set of multiplicity m"

Definition 6.4.2: The receiving set of multiplicity m. The set of all cells that receive synaptic contacts from at least m cells in W is called "the receiving set of multiplicity m." It is denoted by R_m. Any subset of R_m is called "a receiving set of multiplicity m."

From these definitions it is clear that for every $m > 1$,

$$S_{m+1} \subseteq S_m$$
$$R_{m+1} \subseteq R_m$$

Note that there is no requirement that the three sets W, S_m, and R_m be disjoint. It is possible, for instance, to have one cell in W that gives synapses to m cells in W. We have already seen (in Section 6.2.2) that the fact that S_m is the sending set (of multiplicity m) to W does not mean that W is a receiving set (of multiplicity m) of S_m.

Definition 6.4.3: A diverging/converging link. Two sets, W_1 and W_2, for which W_1 is a sending set (of multiplicity m) to W_2, and W_2 is a receiving set (of multiplicity m) of W_1, are called "a diverging/converging link (of multiplicity m)." W_1 is called "the sending node of the link," and W_2 is called "the receiving node of the link."

Definition 6.4.4: A diverging/converging link with no attenuation. A diverging/converging link in which a single spike in the sending node is expected to generate at least one spike in the receiving node is called "a diverging/converging link with no attenuation."

In Section 6.3 we saw that if the ASG of the synapses in a link of multiplicity m is at least $1/m$, then it is a link with no attenuation.

Definition 6.4.5: A diverging/converging chain. A chain of diverging/converging links $L_1, L_2, \ldots, L_n$ (of multiplicity m) for which for every i ($i = 1, 2, \ldots, n - 1$) the sending node in L_{i+1} is the receiving node of L_i is called "a diverging/converging chain (of multiplicity m)." The length of the chain is n.

In Section 6.3 we saw that transmission through a diverging/converging chain is not stable. It either decays to zero or amplifies itself until it reaches saturation. Our contention, therefore, was that transmission through such a chain was most likely to occur by synchronous volleys. This form of transmission is examined in detail in Chapter 7.

6.5 References

Gerald H., Tomlison B. E., and Gibson P. H. (1980). Cell counts in human cerebral cortex in normal adults throughout life using an image analysing computer. *J. Neurol.* 46:113–36.

Griffith J. S. (1963). On the stability of brain-like structures. *Biophys. J.* 3:299–308.

Palmer E. M. (1985). *Graphical Evolution: An Introduction to the Theory of Random Graphs.* Wiley, New York.

7
Synchronous transmission

Our discussion of the transmission of activity through synapses concentrated on asynchronous transmission. We defined the ASG as the number of spikes added to the postsynaptic stream after a presynaptic spike, but we were not concerned with the exact time at which these extra spikes were produced. We have shown that if several spikes converge on one postsynaptic target, their effects add up linearly as long as their EPSPs do not overlap in time.

In this chapter we deal with the properties of synchronous transmission of activity. We have already shown that if a cell receives convergent excitations from several sources, their combined effect will be greater than the sum of their isolated effects. Here we examine this property quantitatively. An examination of the properties of synchronous activation of neurons will lead us to the conclusion that diverging/converging chains are bound to transmit activity synchronously.

The last section deals with the question how one can observe synchronous activity experimentally. In fact, this entire book came about because in studies in which we recorded the activity of several neurons simultaneously, we observed phenomena that could be explained only by assuming synchronous transmission through diverging/converging chains.

7.1 Synchronous gain

When we talk about synchronous transmission through a chain of diverging/converging links, we mean that activity arises after a fixed delay at each node. The requirement of a fixed delay proves to be essential, because only then can we ensure that when several pathways converge onto a single neuron, they will produce synchronous excitation. Therefore, this section deals with transmission from a presynaptic action potential to a postsynaptic action potential with a fixed delay. A

227

postsynaptic spike that follows a volley of synchronized postsynaptic spikes with a fixed delay will be referred to as a "synchronous" spike. The term "fixed delay" requires some clarification. The probability of a response occurring after an absolutely exact delay (say 1.000 . . .) is null. When we speak of nervous activity, there is no point in requiring accuracy that is much higher than the rise time of an EPSP (2–3 ms) or the absolute refractory period (1–2 ms). Thus, a "fixed delay" will be a delay whose variation does not exceed 2 ms.

Let us consider a synchronized volley of presynaptic spikes evoking a large EPSP whose amplitude is equal to the distance between the average membrane potential and the threshold. Whenever such an EPSP occurs, there is a fair chance that it will reach threshold within its rise time and produce a "synchronous" postsynaptic spike. If instead of the regular average membrane potential we use the median, we have a probability of 0.5 of producing a (synchronous) spike within the rise time of the EPSP. Note that in cases in which no action potential is generated during the rising phase of the EPSP, there still is a possibility of generating another action potential during the repolarization phase of the EPSP. The ASG of this EPSP is therefore greater than 0.5. Even the largest EPSP cannot ensure the generation of a synchronous response with absolute certainty, because when it arrives the postsynaptic cell may be deeply hyperpolarized, or even refractory. A single EPSP can produce several spikes, but only the first one can be considered as synchronous. Asynchronous effects add up linearly, but synchronous effects do not. Nevertheless, we wish to give some quantitative measure for the synchronous effects that a given EPSP might have. We can do this by comparing their amplitude to the amplitude of the previously mentioned EPSP (which produces a synchronous response with a probability of 0.5).

Definition 7.1.1: Synchronous gain 50. The ratio between the amplitude (A) of an EPSP and the distance from threshold to the median membrane potential is called the "synchronous gain 50" (SG50).

If the membrane potential fluctuates along a symmetric probability density function (e.g., Gaussian p.d.f.), the median is equal to the average, and we have

$$SG50 = A/T \qquad (7.1.1)$$

where A is the amplitude of the EPSP, and T is the distance from threshold to the average membrane potential.

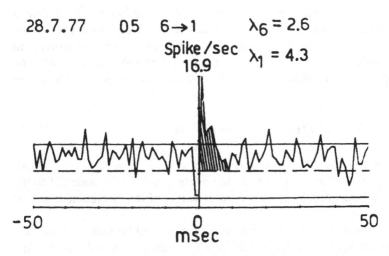

28.7.77   05   6→1 $\lambda_6 = 2.6$

Spike/sec $\lambda_1 = 4.3$
16.9

-50   0   50
             msec

Figure 7.1.1. Cross-correlation function computed from recordings of two cortical neurons. [From M. Abeles, *Studies of Brain Function. Vol. 6: Local Cortical Circuits: An Electrophysiological Study*, 1982, with permission of Springer-Verlag.]

Example 7.1.1. If the threshold of a cell (T) is 10 mV and the amplitude of an EPSP (A) is 0.5 mV, then the SG50 of this synapse is 1/20. This means that it takes the activation of twenty such EPSPs synchronously to generate a synchronous (within the rise time of the EPSPs) action potential in 50 percent of cases.

In general, if the SG50 of a synapse is x, it takes $1/x$ such synapses to produce a synchronous response with a probability of 0.5.

Example 7.1.2. In Figure 7.1.1 we see a cross-correlation function computed from a recording of the activity of two neurons in the auditory cortex of an unanesthetized cat. Let us assume that this curve is due to direct monosynaptic connection between the two cells. We know already how to compute the asynchronous gain of this synapse: We have to integrate the area of the curve above the mean rate of firing. This happens to be 0.06 for this particular case. It means that every spike in the presynaptic train will, on the average, add 0.06 spike to the output train. The timing of these added spikes will be anywhere during the first 10 ms after the firing of the presynaptic cell.

How can we evaluate the SG50 of this same synapse? We can do so if we accept the simple membrane potential fluctuation model developed in Chapter 4. There we assumed that the membrane potential fluctuations were Gaussian and that the firing rate of a cell was proportional to the probability of the membrane potential being above threshold. The proportionality constant was the reciprocal of the refractory period. We obtained

$$\lambda = \{1/[r(2\pi)^{\frac{1}{2}}]\} \int_{T}^{\infty} \exp\{-u^2/2\} \, du \qquad (7.1.2)$$

where r is the refractory period and T is the normalized threshold, expressed as the distance from the average membrane potential to the threshold divided by the standard deviation of the membrane potential fluctuations.

We can now compute the amplitude of the EPSP that generates the response of Figure 7.1.1. We do so by computing the threshold while the neuron is in its resting state and when it is at the peak of the EPSP.

During the resting state, we substitute 4.3 (spikes per second) for λ. By assuming that r is 0.002 s, we can solve equation (7.1.2) for T. This usually will be done by looking at a table of areas under the normalized Gaussian density function. We find that $T = 2.38$. That is, the threshold at rest is 2.38 standard deviations above the average membrane potential.

At the peak of the EPSP, the firing rate (λ) is 16.9 spikes per second. The result for equation (7.1.2) at the peak of the EPSP is $T = 1.83$ standard deviations above the average membrane potential at that instant. This means that at the peak of the EPSP, the membrane potential is depolarized by 0.55 (2.38 − 1.83) standard deviation. The SG50 of this synapse is therefore

SG50 = 0.55/2.38 = 0.23

If we want to reach threshold by synchronous activation of several such synapses, we need about four (1/0.23) of them. If we repeatedly activate these four synapses synchronously, we obtain a synchronous response in 50 percent of the cases.

The asynchronous gain of the same synapse is 0.06. This means that we would have to activate eight (0.5/0.06) such synapses in the asynchronous mode to obtain an average of 0.5 spike asynchronously.

We can generalize this comparison between the synchronous gain and the asynchronous gain by the ratio

coincident advantage = SG50/2ASG

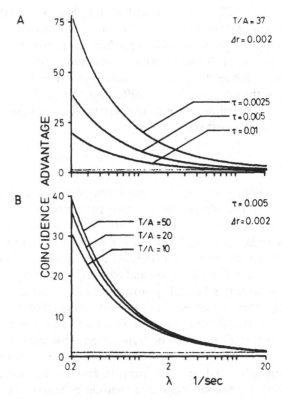

Figure 7.1.2. Advantage of coincident activation of synapses. [From
M. Abeles, "Role of Cortical Neuron: Integrator or Coincidence Detector?"
Israel Journal of Medical Sciences, 18:83–92, 1982, with permission.]

We can calculate the advantage of synchronous activation of the cell by
repeating the analysis of Example 7.1.2 for different synapses under
different conditions. By doing so, we obtain graphs like those in Figure
7.1.2. There, the coincident advantage is plotted as a function of the
spontaneous firing rate of the postsynaptic cell (λ), as a function of the
time constant of recovery of the EPSP (τ), and as a function of the
reciprocal amplitude of the EPSP (T/A). These graphs give us a sense of
the factors that make the neurons sensitive to synchronous activation.
Foremost is the firing rate of the postsynaptic cell, then comes the
duration of the EPSP, and then the size of the EPSP. In the cortex, all
these parameters fall within ranges that make the cells much more sensitive to synchronous activation.

Elsewhere [Abeles, 1982b] the ASG of the average cortical synapse was estimated to be 0.003. A synapse with that asynchronous gain will have an SG50 of about 0.04. This means that it takes about twenty-five such synapses activated synchronously to produce a synchronous output with a probability of 0.5. If the same twenty-five synapses were activated in an asynchronous mode, they would generate only 0.075 of a spike (i.e., one spike in thirteen trials). This difference in sensitivity reflects the way activity is transmitted through converging/diverging links. This issue is examined in the next section.

7.2. Synfire chains

In Chapter 5 we examined ways in which excitation is transmitted through chains of neurons in the cortex. We saw that the simplest concept, in which a burst of spikes in one neuron will excite the next neuron and elicit there a burst of spikes, and so forth, cannot work in the cortex. Our conclusion was that in the cortex, transmission is likely to occur between sets of neurons connected by diverging and converging pathways.

We saw that the large numbers of synapses given by almost all cortical cells to their neighbors guarantee the abundant existence of diverging/converging links, so that there are many potential diverging/converging chains in the cortex. Excitation can be transmitted along such chains in two ways: asynchronously and synchronously. In the asynchronous mode, one set of neurons (the sending node) starts to fire at a high rate. This builds excitation by spatial and temporal summation in the second (receiving) node, which after a while also starts to fire at a higher rate, and so on. In the synchronous mode the chain could operate as follows: Each neuron in the sending node fires one spike, but all the cells in this node fire at the same time. The next node receives a synchronized volley of spikes, which causes all the cells in that node to fire synchronously, and so on. This form of transmission has three major prerequisites:

1. Adequate initial conditions (i.e., all the cells in the sending node must fire synchronously).
2. The synapses must be strong enough to ensure synchronous firing of the cells in the receiving node.
3. The existence of a mechanism that will ensure that small jitter in firing times will not accumulate along the chain to the extent that the firing in each node can no longer be considered synchronous.

Compliance with these three prerequisites will be examined in detail in the following sections. Here we assume that diverging/converging chains can indeed transmit synchronous activity, and we shall attempt to examine the consequences of the greater sensitivity of cortical cells to synchronous activation.

Let us observe a small cortical region of about 0.5 mm^3. In such a region there are about 20,000 neurons that fire at an average rate of five spikes per second. This means that at every millisecond there are about 100 neurons that fire. These neurons fire in synchrony (by our definition of synchrony). Some of these cells are excitatory, and some are inhibitory. We have no information regarding differences in spontaneous firing rates of excitatory and inhibitory cortical neurons. In the cortex, more than two-thirds of the neurons are pyramidal cells that are excitatory. Therefore, we assume that 70 of the 100 synchronous spikes are excitatory. In Section 7.1 we estimated the SG50 of the average cortical excitatory synapse to be approximately 25. By following the computation method of Section 6.4, we find that 70 cells are expected to give diverging/converging connections of multiplicity 25 to about 380 cells. This means that 380 cells receive synchronous activation that is expected to activate them to fire with a probability of 0.5. Thus, in the next millisecond about 190 of these 380 cells are expected to fire. However, because the firing of the cortex is stationary, only 100 cells will fire in the next millisecond. This discrepancy (between the expected firing of 190 cells and the 100 that actually fire) may be due to several factors: the effect of the 30 inhibitory cells, and/or because the calculations of connectivity (from which the figure of 380 was derived) were oversimplified, and/or because the average SG50 of the cortical synapse is somewhat lower than 25.

Although these figures are only crude approximations, they indicate that it is quite possible that whenever a cortical cell fires, it does so because it has been synchronously activated by some cells in its vicinity. This might be considered a trivial outcome of the higher sensitivity of the cortical cell to synchronous excitation, because for any group of cells that happen to fire in a certain millisecond, the cells that will fire in the next millisecond are most likely to be among those that receive convergent synchronous excitation from the previously active cells. As was mentioned in Section 5.4.2, this conjecture has been proved rigorously by Nelken [1988].

Exercise 7.2.1. Assume that for an average cortical neuron (firing at 5/s) the average synaptic ASG is 0.01, and the synaptic transmission curve (Section 3.2) is decaying sequentially with a time constant of 5 ms. What will be the SG50 of the EPSP? How many cells will be synchronously excited by synchronously firing 70 pyramidal cells? Repeat these calculations for an average synaptic ASG of 0.001.

The synchronous activation described in the foregoing paragraphs is of little relevance for information processing, because it cannot yield reproducible patterns of activity. To understand this more thoroughly, let us assume that on another occasion the same 100 neurons happen to fire again, and let us compare the cells that will fire 2 ms later on both occasions. On the second occasion, the 70 excitatory neurons will again activate the same 380 cells and cause them to fire with a probability of 0.5. However, this time a different subset of these 380 neurons will actually fire. This subset, after one additional millisecond, will excite another set of neurons whose overlap with the set of neurons that were excited after 2 ms on the first occasion will generally be small. Thus, if we examine the propagation of activity through the cortical network, we see that although we started on the second occasion with the same 100 cells, within a few milliseconds the activity took a completely different path. This lack of reproducibility is mainly due to the fact that in our example, propagation is carried through a complex network in which all the connections have equal strengths. In this situation the random fluctuations in excitability have a cumulative effect and cause the activity to propagate through different neurons on each trial. The situation can be much more reproducible if within the region there are some diverging/converging connections that are stronger than the average synaptic connection.

In order to illustrate this situation, let us observe a set of 30 neurons that are connected to another set of 30 neurons by a diverging/converging pathway of multiplicity 14 (i.e., each cell in the sending node excites 14 cells in the receiving node, and each cell in the receiving node is excited by 14 cells in the sending node). From Table 6.2.1 we learn that there are numerous such connected nodes in the cortex. However, we assume that the connections within the particular link we are studying are strong and have an SG50 of 1/7. Let us examine what occurs if in a certain millisecond, 15 of the 30 cells from the sending node fire. Each of the cells in the receiving node will receive an average of seven synchronous EPSPs. Because the SG50 of these synapses is 1/7, only half of the receiving cells will fire, so that 15 cells of the receiving node will fire. If the receiving node sends diverging/converging connections to a third node, that will cause 15 cells from the third node to fire synchronously, and so on. If we observe the repeated propagation of activity through this chain, we again see that even if we always start with the same 15 cells in the first node, another subset in the second node and third node will become active on every occasion. However, the 15 active cells always remain confined to the same node of 30 cells. Thus, in this sense synchronous activity can be transmitted repeatedly along a chain of strengthened diverging/converging links. This form of activity will be called "synfire transmission," and a diverging/converging chain that transmits in this mode will be called a "synfire chain." The formal definitions are as follows:

Definition 7.2.1.: Synfire link. A diverging/converging link (Definition 6.4.3) is called a "synfire link" only if there exist n for which the following conditions hold true: (1) Whenever n cells from the sending node become synchronously active, it is expected that at least k cells of the receiving node will become synchronously active. (2) k is not smaller than n.

Definition 7.2.2: Synfire chain. A series of synfire links in which the receiving node of one link is also the sending node of the following link is called a "synfire chain."

7.3 Stability of transmission in synfire chains

Let us again examine the example of a synfire link given in Section 7.2. The link is made up of two nodes, each with 30 neurons. They are connected by a diverging/converging pathway with multiplicity 14. That is, each neuron in the sending node gives excitatory synapses to 14 of the cells in the receiving set, and each cell in the receiving node receives excitatory synapses from 14 cells of the sending node. These synapses have an SG50 strength of 1/7.

In the preceding section we saw that if exactly half of the cells in the sending node fire synchronously, we can expect half of the cells in the receiving node to fire after one synaptic delay. But what will happen if in the receiving node the number of responding cells is slightly more (or less) than half? If we have a long synfire chain, then the receiving node of the link under consideration is also the sending node of the next link. If we want the chain to operate at 50 percent saturation in a stable mode, we should have a situation in which if slightly more than half of the cells in one of the nodes fire, then the expected number of responding cells in the next node will be somewhat lower. On the other hand, if the activity in the sending node is slightly below 50 percent, then the expected numbers of responding cells in the receiving nodes should be slightly higher than the number of cells that fire in the sending node. As we shall see, this is not what actually occurs in a synfire chain. To get some insight into what is expected, let us consider the following example.

Example 7.3.1. In the foregoing synfire link, 20 of the 30 cells in the sending node fire. What is the expected number of synchronously responding cells in the receiving node?

To find the answer, we need to know the amplitude of the EPSPs elicited by the sending neurons in the receiving neurons. Lacking more precise information, we can deduce the amplitude from the following assumptions: (1) The cells fire at an average rate of five spikes per second. (2) The membrane potential of the receiving cells fluctuates with a normal p.d.f. (3) The rate of firing of the receiving cells is proportional to the probability of the membrane potential being above threshold, and the proportionality constant is 500 (the reciprocal of the refractory period).

That the cells fire at five spikes per second implies that the probability of the membrane potential being above the threshold is 0.01 (5/500). For a normally distributed random variable, this probability is achieved when the threshold is 2.33 standard deviations above the mean. The strength (SG50) of the synapses is 1/7 (i.e., seven such EPSPs bring the membrane potential from its mean to the threshold). Therefore, the amplitude of each EPSP is 0.33 (2.33/7) standard deviations.

If 20 neurons in the sending set fire, and each of them produces EPSPs in 14 cells of the receiving set, then these 20 neurons produce 280 (20 · 14) EPSPs, which are distributed among the 30 neurons of the receiving set. On the average, each of the neurons in the receiving node receives 9.33 (280/30) EPSPs, whose combined depolarization is 3.1 (9.33 · 0.33) standard deviations.

During the resting state, the mean membrane potential of the cells in the receiving node is 2.33 standard deviations below threshold. At the peak of the combined EPSP it is 0.77 (3.1 − 2.33) standard deviations above(!) the threshold. Because the area of the normal p.d.f. above this threshold is 0.78, this is the probability that a cell from the receiving mode will fire one synaptic delay after the sending cells fire. Therefore, we expect to see synchronous firing in 23 (0.78 · 30) cells from the receiving set.

To summarize, we have studied a synfire link with the following properties: When 15 cells in the sending node fired synchronously, they were expected to elicit synchronous responses in 15 cells of the receiving node. We found that when 20 cells in the sending node fired synchronously, they were expected to elicit synchronous responses in 23 cells of the receiving node. If we had a long synfire chain made of identical links, it would not transmit activity at a stable level of 50 percent saturation.

Exercise 7.3.1. Using the same method as in Example 7.3.1., compute the expected number of responding cells if only ten of the neurons in the sending node fire.

We can generalize from the computations of the foregoing example and build an input–output curve for the synfire link. We assume that the link has a width of w (i.e., the sending and the receiving nodes contain w cells each) and a multiplicity of m (i.e., each cell in the sending node gives synapses to m cells of the receiving node, and vice versa) and that the synapses have a strength of SG50. We describe the transmission through one link by a curve that shows the expected number of responding cells in the receiving node as a function of the number of synchronously firing cells in the sending node. An example of such an input–output curve is shown by the heavy line in Figure 7.3.1. This curve has several interesting features. Even when there is no activity at all in the sending node, there is some activity (0.005) in the receiving node. This is because of the spontaneous activity (assumed here to be 5/s) of all the cells in the cortex. For the same reason, even when all the cells in the sending node are active, not all cells in the receiving node are expected to respond. Some of them will have fired a fraction earlier and will be refractory. If we have a long chain with identical links, we expect the chain to operate in a steady state in which the activity in the receiving node is equal to the activity in the sending node. The straight line ($y = x$) describes all these points of equilibrium. This equilibrium line crosses the input–output curve at three places: A, B, and C. We can examine the stability of these equilibrium points by considering the evolution of activity when we start the chain at a point that is near equilibrium.

Let us examine what occurs when 60 percent of the neurons in the sending node fire (f_0 in Figure 7.3.1). From the curve, we would expect to elicit a synchronous response in about 66 percent of the cells in the receiving set. The state of this link is represented by the point D_1 on the curve. In the next link, the activity will start with 66 percent of the cells in the sending node being active (point f_1 on the graph), which will be expected to elicit a synchronous response in about 77 percent of the cells

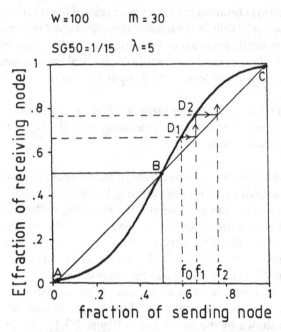

W = 100   m = 30

SG50 = 1/15   λ = 5

Figure 7.3.1. Transmission through a synfire link: the expected response in the receiving node as a function of the activity in the sending node, computed for nodes with 100 cells having diverging/converging connections of multiplicity 30 and an SG50 of 1/15. The cells are assumed to have a spontaneous firing rate of five spikes per second. [From the final examination of I. Warburg.]

in the receiving node of this second link. The state of the second link is represented by D_2 on the graph. If we continue to observe the evolving activity in the third, fourth, and following links, we see that the representing point moves toward equilibrium point C. Similarly, it is easy to see that when the activity of the sending node in the first link is below 50 percent, the activity will gradually drop down as it propagates along the chain to reach point A. This type of analysis shows that of the three equilibrium points A, B, and C, only A and C are stable. Point B represents the *ignition threshold* for the chain. If we start off with an activity level that is below B, the activity will die off as it propagated along the chain, until the cells fire at their normal spontaneous rate. If, on the other hand, we start off with an activity level that is above B, the activity will build up along the chain until it reaches saturation (all the cells will be either firing or refractory).

We must remember that the activity in the chain has a probabilistic nature, so that the input–output curve describes only the expected activity. There will always be fluctuations around this expectation.

Exercise 7.3.2. We have a synfire link, as shown in Figure 7.3.1. How much of the sending node must we activate in order to ensure that the activity in the receiving node will not fall below the ignition threshold in an average of 99 of 100 trials?

In computing the curve of Figure 7.3.1, we assumed the existence of very specific properties:

1. The structure of the chain: It is 100 cells wide, the diverging/converging connections are of multiplicity 30, and the synapses have an SG50 of 1/15.
2. Brain activity: The values of the intracellular membrane potential are distributed normally, the firing rate of the cells is proportional to the probability of the membrane potential being above threshold, with proportionality constant of 500, and the cells fire spontaneously at a rate of five spikes per second.
3. The distribution of partial activation in the receiving nodes: We assume that when only f of the cells in the sending node are active, the mf EPSPs they produce are equally distributed among the receiving cells. In fact, all we know is that each of the receiving cells is expected to be excited by mf/w EPSPs, but the actual excitation will be distributed around this expectation.

Problem 7.3.1. Find a way to express the sensitivity of the transmission along a synfire chain to statistical fluctuations (in activity and in connectivity). Describe how this sensitivity depends on w, m, SG50, and λ.

One way to examine the sensitivity of the synfire transmission to these assumptions is to see what will happen to the input–output curve when the assumptions are altered. This has been done by I. Warburg and some of his results are shown in Figure 7.3.2. By comparing Figure 7.3.1 (background firing rate of 5/s) to Figure 7.3.2A (background firing rate of 1/s) and Figure 7.3.2B (background firing of 20/s), we get a notion of the effect of the spontaneous firing rate of the cells on the transmission. It is clear that the higher the firing rate, the flatter the curve. A flat curve indicates that the evolvement of activity (from threshold to saturation) along the chain is very slow. That, in turn, implies that statistical

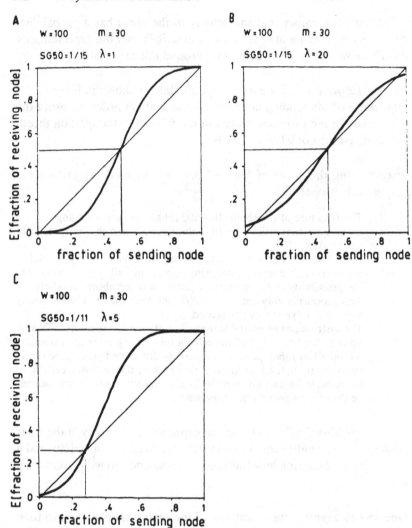

Figure 7.3.2. Synfire transmission under various conditions. All the curves are computed for the same conditions as in Figure 7.3.1, except that one parameter was modified. A: The spontaneous firing rate is 1/s. B: The spontaneous firing rate is 20/s. C: The SG50 is stronger.

fluctuation may quench the activity more easily, even if it is above threshold. It also means that activity in the chain may start off by itself. Perhaps that is what happens when specific memories are awakened by every weak tetanization of a cortical region [Penfield and Rasmussen, 1950].

By comparing Figure 7.3.1 to Figure 7.3.2C, we can also see that the location of the threshold is inversely related to the product $mSG50$. The higher this product, the lower the threshold. When $mSG50 = 2$, the threshold must be at 0.5.

To summarize, synfire chains do not transmit partial activity in a stable fashion. For every chain (composed of identical links), one can define an ignition threshold. If the number of synchronously excited cells in a node is less than this threshold, then as excitation is transmitted from node to node, the activity is likely to fade away. If the number of synchronously excited cells is above this threshold, activity is likely to build up until all the cells in each node that are not refractory will fire synchronously. The level of the ignition threshold depends on the multiplicity of the diverging/converging connections between the nodes and on their synaptic strength. In the cortex, the neurons that take part in one synfire chain are likely to comprise only a small fraction of all the neurons in the region. Therefore, saturated transmission through a synfire chain is expected to have little effect on the total activity of the region.

7.4 Accuracy of timing

When synfire activity propagates along a chain, the degree of synchronization between the spikes may progressively diminish because of several sources of jitter in timing. Even a small amount of desynchronization in each link can accumulate when activity progresses from one link to the next. Such an accumulation of jitter can very well be the limiting factor for the length of the synfire chain. In the following sections, we examine the sources of fluctuations in the delay between a presynaptic spike and the postsynaptic response. In the first section, we study the way in which such a timing fluctuation may build up.

7.4.1 Accumulation of errors

Let us assume that we have a very long diverging/converging chain with strong synaptic connections in which the mean delay of transmission in each link is 1 ms, to which some jitter (say 0.1 ms) is added. As a result, if the cells of the sending node fire with absolute synchrony, the scatter around the mean time of response in the receiving node will be 0.1 ms. Even if we start the activity in this chain with absolute synchrony, we can expect the jitter to build up until after a while it will become longer than 2 ms, and then we can no longer speak of synfire activity in the chain. How long will it take for this to happen?

In a rough approximation, we can build the following model: The transmission time in each link behaves like a random variable having a mean of value of 1 and a variance of 0.01 (0.1^2). The transmission time after N such links behaves like the sum of N independent random variables. That is, the mean of the sum is N ms, and the variance of the sum is $0.01N$. As long as the standard deviation does not exceed 2 ms (i.e., the variance is 4 ms²), we can consider transmission to be synchronous. This limit is reached only after 400 links (4/0.01).

This model is oversimplified for several reasons. The transmission jitter is due to two types of processes: (1) differences in the conduction times of the axons leading the spikes from the sending node to the receiving node; (2) variability in the time of generation of a postsynaptic spike caused by the random nature of the transmitter release at each synapse and the random nature of the level of the membrane potential when the EPSPs arrive. These two sources of variability have different rules for accumulation.

In addition to mechanisms that generate jitter, the synfire chain is equipped with mechanisms to reduce the jitter, as described later. Each cell in the receiving node "sees" a volley of spikes. Its response will depend on the mode of arrival times of the spikes within this volley, not on the timing for each individual spike within the volley. The mode has a variance that is smaller than the variance of the arrival times of the individual spikes.

The threshold of the chain's ignition is lower than its width. Therefore, if the trailing spikes in the volley arrive too late, they may be ignored. The early spikes may be sufficient to elicit activity in most of the cells in the receiving set, and this may be enough to ensure synchronous transmission downstream.

Thus, the diverging/converging connections tend to resynchronize the activity within a node. We can gain some insight into this resynchronization by examining the following case:

Example 7.4.1. (Proposed by I. Nelken as part of his final examination). Suppose we have a complete Griffith chain that is w cells wide.

That is, each node is made of w cells, and each of them is connected to every cell in the next node. Suppose, also, that there are no fluctuations in the membrane potentials, in the conduction times, or in the synaptic delays. Let us further assume that all the cells and connections are homogeneous. That is, the membrane potentials and thresholds of all the cells are the same, as are the conduction times from any cell in the sending node to every cell in the receiving node. By these assumptions we eliminate all the sources that may add jitter to the firing times of the cells in the receiving node.

Assume that the firing times of the cells in the sending node are somewhat dispersed. What will be the dispersion of firing times of the cells in the receiving node?

Let us denote the synaptic strength of the link by SG50 (if the link is functional, then $1/SG50 < w/2$). Each cell in the receiving node requires synchronous activation by $1/SG50$ cells from the sending node in order to reach threshold. Because the sending cells are not activated in complete synchrony, the first EPSPs that arrive may start to repolarize by the time that the $1/SG50$th EPSP arrives, so that the cell will reach threshold only during the rising phase of the $1/SG50 + 1$st EPSP. However, we assume that all the cells in the receiving node are at the same state, so that all of them will fire at the same point along the $1/SG50 + 1$st EPSP. Thus, even though the activities in the cells of the sending node are not completely synchronized, the elicited activities of all the cells in the receiving node are completely synchronized.

The dispersion of activity in the sending node will be expressed in the delay between the mean time of activation of the sending node and the synchronous response of the receiving node. This can be seen from the following discussion. Let us assume that we repeatedly activate the sending node (with the same statistical dispersion). The receiving node will respond every time that the $1/SG50 + 1$st spike arrives. The arrival time of this spike will change from activation to activation, and therefore the delay until the receiving node responds will vary.

Note that the dispersion of delays will be smaller than the dispersion of firing times in the sending node, because waiting for the $1/SG50 + 1$st event to occur is a form of averaging. In any case, though the delay until the receiving node responds is subject to random fluctuations, the response itself consists of a completely synchronized volley of all the cells in the node.

All these reasons suggest that the jitter builds up at a rate slower than the $\sigma\sqrt{N}$ law suggested earlier.

Problem 7.4.1. Assume that the firing times in the sending node are distributed normally and that the cells behave ergodically (i.e., the statistics describing the times of activity of one cell in repeated trials are the same as the statistics for the entire population on a single trial). Assume further that the transmission time from the sending node to the receiving node is fixed. What will the distribution of response times in the receiving node look like?

7.4.2 Jitter due to conduction time

In this section we estimate the effects of conduction time from the cells in the sending node to the cells in the receiving node. Let us assume that the cells of the two nodes are distributed in a small cortical volume, so that the distance from a cell in the sending node to a cell in the receiving node may be as short as zero or as long as 250 μm. Even if we assume that the distances are distributed evenly between these two extremes, we get an average distance of 125 μm and a standard deviation of 72 μm ($250/\sqrt{12}$). If the velocity of propagation along intracortical axons is about 2 m/s [Burns, 1958], then the standard deviation of the conduction time is 36 μs. How will that dispersion of arrival times affect the EPSPs in the cells of the receiving set?

If every fiber generates a fixed EPSP whose shape is EPSP(t), and if the arrival times of many (m) such EPSPs are distributed with a p.d.f. of $f(t)$, then the resulting EPSP is given by the convolution of the p.d.f. and the original EPSP:

$$E(t) = m\text{EPSP}(t)*f(t) \tag{7.4.1}$$

where $E(t)$ is the resulting EPSP, and the asterisk denotes convolution.

Exercise 7.4.1. Show that if we repeatedly excite the same presynaptic fiber and measure the average EPSP in the postsynaptic cell, then the expectation of the EPSP is

$$E[\text{EPSP}(t)] = f(t)*\text{EPSP}(t)$$

where $f(t)$ is the p.d.f. of the delays from the time of stimulation to the start of the EPSP, and the asterisk denotes convolution.

M converging EPSPs are expected to generate strong depolarization in the postsynaptic cell. The main factor that will control the response time in a very large EPSP is its rise time. If the rise time is 1–2 ms and the

EPSP's rising phase is close to linear most of the time, then a narrow distribution of arrival times will have little effect on most of the rise time. This can be seen by solving the following exercise:

Exercise 7.4.2. Compute the convolution between the ramp

$$\text{EPSP}(t) = \begin{cases} 0 & \text{for } t < 0 \\ t & \text{for } 0 < t \leq 1 \\ 1 & \text{for } 1 < t \end{cases}$$

and the rectangular distribution

$$f(t) = \begin{cases} 0 & \text{for } t \leq 0 \\ 10 & \text{for } 0 < t \leq 0.1 \\ 0 & \text{for } 0.1 < t \end{cases}$$

It seems that under these conditions the variation of conduction times from the sending node to the receiving node has only negligible effects on the response times of the cells in the receiving set.

The situation may be very different when action potentials have to traverse long distances before they arrive at the receiving set. It is conceivable that synchronous transmission is not possible between the thalamus and the cortex, or even between two cortical regions. One may suggest that the columnar mode of wiring from thalamus to cortex, as well as from cortex to cortex [Goldman and Nauta, 1977], is useful because it reduces the jitter in the conducting distances between sending and receiving structures.

7.4.3 Jitter due to the synaptic processes

An additional source of fluctuations is the random nature of the transmitter release at the presynaptic endings. Both the amount of transmitter released (quantal content) and its time may fluctuate from action potential to action potential. In the synfire chain, we are dealing with relatively large summed EPSPs. Therefore, the changes due to fluctuations in quantal content of transmitter presumably are small.

As yet we do not have any direct data about jitter in the release time of the corticocortical synapse. Measurements made for the Ia fibers in the spinal cord [Cope and Mendell, 1982] show that the standard deviation for transmission between these fibers and a motoneuron is about 50 μs.

For synchronous transmission, it is the total arriving volley that mat-

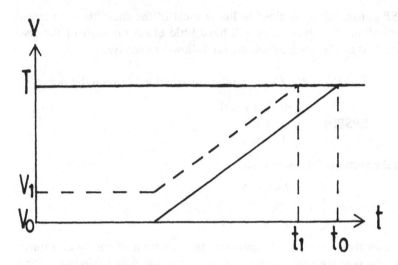

Figure 7.4.1. Fluctuations in response time due to fluctuations of the membrane potential.

ters. If the volley consists of m spikes, the fluctuations in quantal content and mean arrival time of the volley will be smaller by $\sqrt{m}$.

7.4.4 Effects of membrane potential fluctuations on response time

When activity is transmitted in a synfire chain, every cell is excited by a very strong depolarization, which is generated by the summation of many synchronous EPSPs. In this case the membrane potential usually will reach threshold during the rising phase of the summated EPSP. Even if the shape of this summated EPSP is not subject to any noise, it will be superimposed on a membrane potential that fluctuates. Therefore, on repeated trials, the time that elapses until the threshold is reached will vary. In order to reach a deeper understanding of this process, let us examine a simple case in which the EPSP rises linearly, and the rise time is short compared with the time constant of the membrane fluctuations.

In Figure 7.4.1 we see the rising phase of an EPSP. The firing time of this neuron is determined by the time at which the EPSP crosses the threshold. If at the foot of the EPSP the membrane potential is at v_0 (solid line in Figure 7.4.1), the response time will be at t_0. If at the foot of the EPSP the membrane potential is at v_1 (broken line in Figure 7.4.1), the response time will be at t_1. The ratio between the jitter in membrane potential and the jitter in response time is the slope of the EPSP (a):

$$(v_1 - v_0)/(t_1 - t_0) = a \qquad (7.4.2)$$

If the cell's membrane potential has an average value of v_0 and it fluctuates around this value with a standard deviation of σ_v, then the mean delay of the response will be t_0 and it will fluctuate with a standard deviation σ_t, which will be related to the standard deviation of the membrane potential fluctuation by

$$\sigma_t = \sigma_v/a \qquad (7.4.3)$$

Let us now examine our model of a cortical neuron (from Section 7.1) and evaluate the jitter in timing caused by the membrane potential fluctuations. The cell fires at a rate of five spikes per second, meaning that its threshold is $2.33\sigma_v$ from the mean membrane potential. If the cell is embedded in a synfire chain similar to the one described in Section 7.3, then the peak depolarization when all the cells in the giving set fire will be aproximately twice this distance. Thus, the amplitude of the combined EPSP will be $4.66\sigma_v$. The EPSP rises to this height in 1–2 ms. If the rise were linear, its slope would be at least

$$a = 4.66\sigma_v/0.002 = 2330\sigma_v/s$$

According to equation (7.4.3), the standard deviation of the response time will be

$$\sigma_t = \sigma_v/(2330\sigma_v/s) = 0.00043 \text{ s}$$

which is 0.43 ms.

Although the real jitter will be somewhat smaller (because the initial slope of the EPSP is higher than its average slope), this calculation indicates that the jitter due to membrane potential fluctuation is much higher than the jitter from all the other sources.

Problem 7.4.2. How will the jitter in response time relate to the jitter in membrane potential if we relax the requirement that the rise time of the EPSP be much shorter than the membrane potential time constant?

Let us reiterate that in a synfire chain it is the shape of the total volley that matters. Even if the response time of each cell fluctuates with a

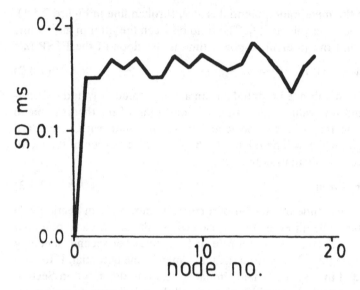

Figure 7.4.2. Buildup of jitter along a synfire chain. The first node in a diverging/converging chain is activated with absolute synchrony. The standard deviation of firing times in each subsequent node is plotted as a function of the position of the node in the chain. [From S. Pantilat, "Theoretical Model of Nervous System," Ph.D. thesis, 1986, The Hebrew University, Jerusalem, with permission.]

standard deviation of 0.43 ms, the fluctuation of the mean time of arrival of the volley in the next node will be much smaller. For instance, if the multiplicity of the connections is 15, the standard deviation for the arrival time of the whole volley will be $0.43/\sqrt{15} = 0.111$ ms. It is clear from these examples that the synchronous transmission through a diverging/converging chain will be more resistant to statistical fluctuations if the multiplicity of the connections is larger and if the strength of the synapses is greater.

One can study the balance between the desynchronizing effects described thus far and the inherent tendency of a diverging/converging link to transmit only the synchronized portion of the incoming volley through simulation, and Figure 7.4.2 shows such a simulation. In this simulation, a diverging/converging chain (composed of 20 nodes, each of which is 10 cells wide and in which the multiplicity of connections is 8) is embedded in a large network of randomly firing nerve cells. The first node of the chain is stimulated synchronously, and the propagation of activity along the chain is observed. For each node, the standard deviation of firing times of the cells in the node is calculated. The chain is activated 100 times, and the standard deviations for each node are averaged. As seen in Figure 7.4.2, the firing times at the second node have a standard deviation of approximately 0.15 ms. The jitter builds up slightly more in the next two nodes, but remains essentially constant thereafter.

7.4.5 Summary

When we are dealing with the transmission of synchronous activity across short distances, the fluctuations of membrane potential are the most significant sources of timing jitter. They can cause fluctuations of about 0.5 ms in the firing time of each individual neuron in the receiving node. However, in a synfire chain, what is significant is the arrival time of a whole volley. The jitter of arrival times of such volleys is in the range of 0.1–0.2 ms.

The sensitivity of the cortical neuron to simultaneous excitation tends to resynchronize a dispersed volley at each node. This tendency balances the random fluctuations in transmission time, so that activity may propagate through many links, with a small but stable jitter.

When activity passes from one cortical region to another, the jitter in conduction times may become the major source of desynchronization. In such cases there must be additional factors that allow synchronous transmission. It is possible that synchronous transmission is carried only locally, whereas transmission across large distances is carried by asynchronous volleys.

7.5 Conversion from synchronous activity to asynchronous activity, and vice versa

The relevant aspect of firing in the peripheral nervous system is, in most cases, the rate of firing of individual neurons, rather than the coordination of firing times between different neurons. For instance, the force of contraction of a muscle depends on the firing rates of the motoneurons that innervate it. In the motor cortex as well, one finds that the firing rate of the neurons is directly correlated with the force, the velocity, and the direction of the movement [Evarts, 1968; Georgopoulos et al., 1989]. Yet at the same time, one finds that some of the neighboring cells in the motor cortex tend to fire with strict synchrony when the appropriate muscle is activated [Allum, Hepp-Reymond, and Gysin, 1982].

If information processing in the cortex is carried out by synfire chains, then the synchronous activity must be translated into bursts of firing at high frequency before it can appropriately affect brain stem and spinal cord neurons. One can easily see how a nerve cell translates a single synchronous volley of excitatory inputs into a burst of spikes. When the synchronous volley arrives, it generates a very large EPSP in the postsynaptic (receiving) cell. Hitherto we have dealt with the first spike generated by this volley and investigated the level of synchronization with the activity of the other cells in the same receiving node. Now we wish to study what occurs after the firing of the first spike. We dealt with this question in Chapter 4, when we discussed the effect of large EPSPs on the firing rate. There we observed that two types of behaviors can be expected. If the cell has strong after-hyperpolarization, and if the exciting synapses are located close to the cell body, then the first action potential erases all the effects of ESPSs that arrived previously. This type of neuron will fire again only when enough new excitation arrives. The motoneurons of the spinal cord show this kind of behavior. However, if the cell does not have significant after-potentials, and if much of its excitation derives from remote dendrites, or if the dendrites can actively generate depolarizing currents [Llinas and Yarom, 1981], then the depolarized dendrites continue to withdraw current from the cell body even after it fires an action potential. In such a cell, a large EPSP generates a burst of spikes.

In the context of a synfire chain composed of cells of the latter type, we can expect that the cells will generate bursts of spikes when excited by a synchronous volley of EPSPs. In a given receiving node, all the cells may burst. The first spike in the burst of each cell is synchronized with the first spikes of the other cells, and the remainder of the spikes in the burst may or may not be synchronized. The amount and rate of firing in each burst will depend on the strength of the cell's activation, as well as its morphology and electrophysiological properties. One can imagine that the cells in the last link of a synfire chain may be specialized in converting the incoming synchronous volley into a burst of spikes.

The inverse problem exists at the sensory end of the nervous system. When a complex stimulus pattern falls on the sensory receptive surface (a picture on the retina, an accord in the cochlea, or a combination of positions of joints during a certain body posture), we think that information regarding the nature of the stimulus is coded in the combination of sensory fibers that fire at a high rate. There need not be and indeed usually are no precise time relations among the firings in different sensory fibers. Is there a simple way by which such an elevated, but not synchronized, level of activity in a population of cells can be converted into synchronous activity?

The simplest way to convert asynchronous activity into synchronous activity is to utilize the increased sensitivity of the diverging/converging chain to synchronous excitation. Let us assume that the first node of such a chain is composed of cells that represent a particular complex stimulus. Whenever this stimulus is presented to the animal, these cells fire at high rates, but without any internal synchronization among the cells. When firing at a higher rate, several of the cells may fire together by mere chance. If the first links in the chain are strong enough, this occasional synchronous firing of some of the cells may be above the threshold for synchronous activation of the receiving node. This means that the number of synchronously firing cells in the next node will be higher. Transmission through the next few links will further increase the degree of synchronicity of the transmitted activity, so that after a few links all that propagates is synfire activity. The next example examines the feasibility of this mechanism.

Example 7.5.1. A certain external stimulus activates 100 neurons that constitute the first node of a diverging/converging link. Every one of these cells is excited by the stimulus for 40 ms and fires during that period at a rate of 200/s. The threshold for setting synfire activity in the chain is 25 (out of 100) cells that fire within 1 ms. What is the probability that this external stimulus will set off synfire activity in the chain?

To simplify the problem, let us divide time into forty sections of 1 ms each. Let us first calculate the probability of obtaining 25 or more cells activated in the same millsecond. For each cell, the probability of firing in a given millisecond is

$$p = \lambda \Delta t = 200 \cdot 10^{-3} = 0.2$$

Accordingly, its probability of not firing during a given millisecond is

$$q = 1 - p = 0.8$$

The probability that any n cells will fire in a given millisecond is obtained by the binomial distribution:

$$p_n = \binom{100}{n} p^n q^{100-n}$$

In our example, $Npq = 100 \cdot 0.2 \cdot 0.8 \doteq 16 >> 1$, and we can therefore use the normal approximation to the binomial distribution with an average of $Np = 20$ and variance of $Npq = 16$. The probability that 25 or more cells will fire in the same millisecond is given by

$$P_{25} = \sum_{n=25}^{100} [1/(2\pi 16)^{\frac{1}{2}}]\exp\{-(n-20)^2/8\} = 0.11$$

Therefore, within these 40 ms we can expect that 25 or more cells will be excited synchronously $x = 40 \cdot 0.11 = 4.4$ times. The probability of not having even one instance in which 25 or more spikes are excited synchronously, when the expected number of times is 4.4, is given approximately by the Poisson distribution:

$$\mathrm{pr}\{n; x\} = e^{-x}x^n/n!$$

which in our case yields

$$\mathrm{pr}\{failure\} = \mathrm{pr}\{0; 4.4\} = e^{-4.4} \approx 0.01$$

Thus, the first node of our chain will be excited above threshold at least once with a probability of 0.99.

This example uses ordinary physiological values. It shows that it is indeed possible to set off synfire activity in a diverging/converging chain by simply exciting its first node at a higher rate for a while. This conjecture was confirmed in a stimulation of such a chain embedded in a randomly connected network [Pantilat, 1986].

Example 7.5.1 also illustrates the following: Let us assume that we study the response properties of one of the cells in the first node by constructing a PSTH (peristimulus time histogram) of its activity around the stimulus onset. We see that it fires an average of eight extra spikes ($200 \cdot 0.040$) after each stimulus, and these eight spikes are evenly distributed during a period of 40 ms. If we construct a PSTH for one of the cells down the chain, we find that it fires only an extra 4.4 spikes per stimulus. We may conclude that the responses of the cells are attenuated along the chain, whereas in fact (if we consider synchronism as the relevant parameter) they have been greatly amplified. Figure 7.5.1 illustrates these points. It is based on a stimulation of a diverging/converging chain that was 20 nodes long and 10 cells wide, in which the neurons of the first node were stimulated asynchronously for 20 ms.

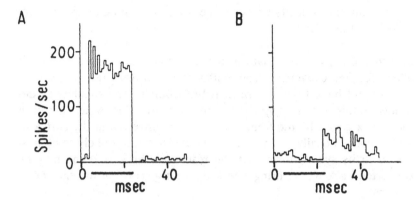

Figure 7.5.1. PSTHs for the responses of a cell in the first node (A) and a cell in the last node (B) of a diverging/converging chain. The chain was activated by exciting the 10 cells in the first node at random times. Every presentation of the stimulus ignited the synfire chain at least once. The responses of the cells in the last node were always strictly synchronized among each other; however, a PSTH for a cell in this node gives the (false) impression of weak responses. [From S. Pantilat, "Theoretical Model of Nervous System," Ph.D. thesis, 1986, The Hebrew University, Jerusalem, with permission.]

Exercise 7.5.1. For the chain described in Example 7.5.1, what should be the SG50 of the synapses if we wish to ensure that the ignition threshold is 25 (assume that the diverging/converging connections are of multiplicity 30)?

Exercise 7.5.2. Assume that we have a diverging/converging chain 25 cells wide. All 25 cells of the first node are excited, by an external stimulus, to fire at a rate of 200 spikes per second for 40 ms. What should be the ignition threshold of the chain if we wish to ensure that this stimulus will set off synfire activity in the chain at least once with a probability of 0.99?

Problem 7.5.1. In Example 7.5.1 we divided the time into forty nonoverlapping time sections of 1 ms. This is rather artificial. For instance, it is quite possible that in one such section there will be only 20 firing cells, and in the next section there will also be only 20 firing cells, but if we could see in between these two sections, we would find an intermediate 1-ms section in which 25 cells had fired.

How can this difficulty be overcome while calculating the probability of igniting the chain?

To summarize, the very nature of the cortical neuron is such that it allows for free conversion from bursts of a high firing rate into synfire activity and back from synfire activity to bursts. The simulations described in this section suggest that sensory information, which arrives in the cortex coded by the firing rate code, is converted in the cortex to synfire code. Locally, in the cortex, the information is likely to be processed in its synfire form, but the results of this processing may be converted back to the firing rate code for transmission to other brain regions.

7.6 Detectability of synfire activity in the cortex

The most direct way to detect synfire activity is to record from several cells in one node and from several cells in another node downstream, and to show that whenever the first cells fire together, then after a fixed delay the other cells also tend to fire together. However, at present, the chances of obtaining such data experimentally are extremely low. This can be seen by considering the following example:

Example 7.6.1. The cells of one node in a synfire chain are distributed in a small cortical volume of 0.5 mm^3. In a cat (or monkey) this volume contains about 5,000 neurons. If the node is 30 cells wide, the probability of recording a cell from this node is $p = 0.006$ (30/5,000). The probability of recording two cells from this node is $p^2 = 0.000036$, and so forth.

The chances of finding two (or more) cells from the same node are improved if the cells in the node are distributed in a smaller cortical volume. Thus, in the visual cortex [Toyama, Kimura, and Tanaka, 1981; Michalsky et al., 1983], in the motor cortex [Allum et al., 1982], in the auditory cortex [Bloom, 1986], and in the brain stem [Frostig, Frostig, and Harper, 1984; Madden and Remmers, 1982] it was found that on several occasions pairs of adjacent neurons tended to fire together within less than 1 ms. However, these observations are not sufficient to indicate the existence of synfire chains, because two neighboring neurons in the cortex could fire together merely because both of them are excited by a strong input from a single thalamic fiber.

The synfire chain is unique in its ability to transmit synchronous activity across many links with very little dispersion. As was shown in Chapter 3, it is not possible to obtain accurate timing after long delays by simple spatial and temporal summation of asynchronous activity.

A simple way to attempt to reveal activity that is specific to synfire chains is to record from two cells of the same chain and look for cases in which firing in one cell is, after a long delay, tightly coupled to firing in the other cell. This approach requires that we identify two cells from the same chain. Let us evaluate the chances of sampling two cells from the same chain by considering the following example:

Example 7.6.2. Let us assume that the chain is embedded in a cortical column $0.5 \times 0.5 \times 2$ mm and that it is 30 cells wide and 100 nodes long. This cortical volume contains about 20,000 cells, and the chain contains 3,000 cells. The probability of obtaining one cell from the chain is $p = 3,000/20,000 = 0.15$, and the probability of obtaining two cells from the same chain is $p^2 = 0.023$. This is not as low as it might seem, because it is possible to record activities simultaneously from several neurons; see Gerstein et al., [1983] and Kruger [1983] for recent reviews. When recording from 10 neurons, one can expect to find an average of 1.5 neurons from the chain. The probability of finding two or more neurons from the same chain when 1.5 are expected may be computed again from the Poisson distribution:

$$\sum_{i=2}^{10} \text{pr}\{i; 1.5\} \simeq 1 - \text{pr}\{0; 1.5\} - \text{pr}\{1; 1.5\} = 1 - e^{-1.5} - 1.5e^{-1.5} = 0.44$$

So it is possible to obtain the desired recording about once for every two trials.

However, even when we record from two cells of the same chain, it may be difficult to distinguish between random activity and activity associated with the synfire activity. This can be seen by considering the following example.

Example 7.6.3. Let us assume that we are measuring the activities of two cells from the same synfire chain. Both cells are spontaneously active at a rate of five spikes per second. Occasionally synfire activity sweeps through the chain; then one of the cells fires, and after T (± 1) ms the other cell also fires. Let us assume that during 100 s of recording, the chain is activated five times. Can we detect the synfire activity in these conditions?

To reveal the synfire activity, we have to construct a cross-correlation function between the two cells with a bin width of 2 ms. The average count in each bin (due to the random spontaneous firing) is expected to be

$$x = T\lambda_1\lambda_2\Delta t = 5 \cdot 5 \cdot 100 \cdot 0.002 = 5$$

Thus, if the two neurons fire independently, the histogram will contain, on the average, five counts in each bin. The actual counts will fluctuate around this average according to the Poissonian probability $\mathrm{pr}\{n; 5\}$. In the bin around the delay T, we obtain an additional five counts due to the synfire activity. Thus, we obtain approximately ten counts at T. The difference between the expected five for any bin and the ten counts in the bin around T is barely significant (the probability of obtaining ten or more counts when five counts are expected is approximately 0.03).

This example shows that the detection of synfire activity becomes feasible when the expected number of cases in which the cells fire at the preferred delay by chance (x) is reduced, or when the frequency of activation of the synfire chain is increased. This means that it will be easier to detect synfire activity among cells that fire spontaneously at low rates. It also means that if the expected synchrony is very good, the histograms should be constructed with the smallest conceivable bin (i.e., a bin that is of the same size as the expected variance in synchronism).

Detectability of synfire activity can be improved if the chain under study is frequently activated. A high rate of activation of the synfire chain can be obtained in awake animals, in which the same pattern of behavior is elicited repeatedly.

If we are fortunate enough to record more than two cells from the same synfire chain, then the difference between the expected random count and the counts due to synfire activity becomes much greater. This can be seen by comparing Examples 7.6.3 and 7.6.4.

Example 7.6.4. Suppose that under the same conditions as in Example 7.6.3 we are able to record from three cells of the same synfire chain. We can attempt to detect the synfire activity by constructing a threefold correlation histogram for the three spikes. The number of expected counts in one bin of such a histogram is [Abeles, 1983]

$$x = T\lambda_1\lambda_2\lambda_3\Delta t^2/2$$

In our example the expected number of counts will be

$$x = 100 \cdot 5 \cdot 5 \cdot 5 \cdot 0.002^2/2 = 0.025$$

If (as in Example 7.6.3) the synfire chain is activated five times during the 100 s of recording, we obtain five counts in the bin that describes the appropriate combination of delays. There is an extremely low probability of observing five counts by chance when the expected count is 0.025. Thus, time-locking among three neurons is a more sensitive method of detecting synfire chain activity than time-locking between two neurons. An efficient algorithm for detecting all patterns that appear repeatedly in a recording from several parallel spike trains has recently been described [Abeles and Gerstein, 1988].

This example illustrates that in order to detect activity associated with synfire chains, one should try to record several neurons in the chain. This can be achieved by simultaneously recording the activities of as many single units as possible.

There have been several published reports of signs of time-locking among several spikes. Dayhoff and Gerstein [1983] studied intervals between successive spikes of a single unit and found that sequences of up to seven successive intervals tended to appear at rates that could not be explained by chance. Abeles [1982a] and Abeles et al. [1983] found three-spike sequences with delays of up to 500 ms in the auditory cortex. Frostig et al. [1985] found preferred sequences of up to nine spikes in the medial forebrain cortex and in the amygdala. Landolt, Renis, and Weiss [1985] found such patterns in multiunit activity recorded from the visual cortex, as did Legendy and Salcman [1985].

In the complex firing patterns that were detected in these experiments, a given neuron could contribute several spikes to a given pattern. This is in accordance with the property of synfire chains that allows a given neuron to participate in more than one node.

To summarize, the most plausible way of detecting synfire activity in the cortex is to record from several nerve cells belonging to the same chain. Then, synfire activity is manifested in the accurate time relations among the spikes of these neurons. The chances of pinpointing these spikes (which are time-locked to each other) from the ongoing activity are increased if the synfire activity is elicited frequently and if the time-locking is accurate. The chances of obtaining several nerve cells from the same chain are improved if the activities of many neurons are recorded simultaneously.

In conclusion, we should mention that synfire chains are equivalent to multilayered perceptrons (Section 5.6) in the following sense: Consider a cortical region in which multiple synfire chains are embedded. Call the neurons in the first stage of all the synfire chains the input layer of a multilayered perceptron; call all the neurons in the second stage of all the synfire chains the first layer of hidden units, and so forth. It is evident that the collection of synfire chains becomes a multilayered perceptron. However, in this case, the parameter to be measured is not which neurons (at a given layer) fire at high rates but which neurons (at a given layer) fire synchronously. We should note that synfire chains seem to bridge the anatomical–physiological discrepancy described in Chapter 5. They can be formed in a network of multiple (feedback) connections, but they operate as a feed-forward network.

7.7 References

Abeles M. (1982a). *Studies of Brain Function. Vol. 6: Local Cortical Circuits: An Electrophysiological Study*, pp. 28–30. Springer-Verlag, Berlin.
    (1982b). Role of cortical neuron: Integrator or coincidence detector? *Isr. J. Med. Sci.* 18:83–92.
    (1983). The quantification and graphic display of correlations among three spike trains. *IEEE Trans. Biomed. Eng.* BME-30:235–9.
Abeles M., De Ribaupierre F., and De Ribaupierre Y. (1983). Detection of single unit responses which are loosely time-locked to a stimulus. *IEEE Trans. Syst. Man Cyber.* SMC-13:683–91.
Abeles M. and Gerstein G. L. G. (1988). Detecting spatio-temporal firing patterns among simultaneously recorded single neurons. *J. Neurophysiol.* 60:909–24.
Allum J. H. J., Hepp-Raymond M. C., and Gysin R. (1982). Cross-correlation analysis of interneuronal connectivity in the motor cortex of the monkey. *Brain Res.* 231:325–34.
Bloom M. J. (1986). Neuronal interactions in the auditory cortex of the cat during binaural stimulation of sound movement. Ph.D. thesis, University of Pennsylvania, Philadelphia.
Burns B. D. (1958). *The Mammalian Cerebral Cortex. Monographs of the Physiological Society, Vol. 8.* Arnold, London.
Cope T. C. and Mendell L. M. (1982). Parallel fluctuations of EPSP amplitude and rise time with latency at a single Ia fiber-motoneuron connection in the cat. *J. Neurophysiol.* 47:455–68.
Dayhoff J. E. and Gerstein G. L. (1983). Favored patterns in spike trains. II. Application. *J. Neurophysiol.* 49:1349–63.

Evarts E. V. (1968). Relation of pyramidal tract activity to force exerted during voluntary movement. *J. Neurophysiol.* 31:14–27.

Frostig R., Frostig Z., and Harper R. (1984). Information trains. The technique and its use in spike train and network analysis, with examples taken from nucleus parabrachialis medialis during sleep–waking cycles. *Brain Res.* 322:67–74.

Frostig R. D., Frostig Z., Frysinger R. C., Schechtman V. L., and Harper R. M. (1985). Multineuron analysis reveals complex patterns of interaction among neurons in forebrain networks and cardiorespiratory parameters during sleep-waking states. *Soc. Neurosci. Abstr.* 11:1010.

Georgopoulos A. P., Lurito J. T., Petrides M., Schwartz A. B., and Massey I. T. (1989). Mental rotation of the neuronal population vector. *Science* 243:234–6.

Gerstein G. L., Bloom M. J., Espinosa I. E., Evanzuk I. E., and Turner M. R. (1983). Design of a laboratory for multineuron studies. *IEEE Trans. Syst. Man Cyber.* SMC-13:668–676.

Goldman P. S. and Nauta W. J. H. (1977). Columnar distribution of cortico-cortical fibers in the frontal association limbic and motor cortex of the developing rhesus monkey. *Brain Res.* 122:393–413.

Kruger J. (1983). Simultaneous individual recordings from many cerebral neurons: Techniques and results. *Rev. Physiol. Biochem. Pharmacol.* 98:177–233.

Landolt J. P., Reinis S., and Weiss D. S. (1985). Identification of local neuronal circuits in the visual cortex of the cat. *Soc. Neurosci. Abstr.* 11:1010.

Legendy C. R. and Salcman M. (1985). Bursts and recurrence of bursts in the spike trains of spontaneously active striate cortex neurons. *J. Neurophysiol.* 53:926–39.

Llinas R. and Yarom Y. (1981). Electrophysiology of mammalian inferior olivary neuron in vitro. Different types of voltage-dependent ionic conductances. *J. Physiol. (Lond.)* 315:549–67.

Madden K. P. and Remmers J. E. (1982). Short time scale correlations between spike activity of neighbouring respiratory neurons in nucleus tractus solitarius. *J. Neurophysiol.* 48:749–60.

Michalsky M., Gerstein G. L., Czakowska J., and Tarnecki R. (1983). Interactions between cat striate cortex neurons. *Exp. Brain Res.* 51:97–107.

Nelken I. (1988). Analysis of the activity of single neurons in stochastic settings. *Biol. Cybern.* 59:201–15.

Pantilat S. (1986). Theoretical model of nervous system. Ph.D. thesis, The Hebrew University, Jerusalem.

Penfield W. and Rasmussen T. (1950). *The Cerebral Cortex of Man: A Clinical Study of Localization of Function.* Macmillan, New York.

Toyama K., Kimura M., and Tanaka K. (1981). Organisation of cat visual cortex as investigated by cross-correlation technique. *J. Neurophysiol.* 46:202–14.

Appendix: Answers and hints

Exercise 1.5.1. Let us denote the probability density function (p.d.f.) of the radii by $f(r)$. Each small range of radii $[r, r + dr]$ may be found with a probability of $f(r)dr$. The average radius is given by

$$\bar{r} = \int_0^\infty rf(r)\,dr$$

The corrected thickness W_e is now a function of r. Equation (1.5.1) reads

$$W_e(r) = W + 2r \quad \text{and} \quad \frac{d}{dr}W_e(r) = 2$$

Therefore, the p.d.f. of $W_e(r)$ is $0.5f(r)$ for $W_e(r) \geq W$, and zero otherwise. The average corrected thickness is given by

$$\overline{W_e} = \int_W^\infty W_e(r) \cdot 0.5f(r)\,dW_e(r)$$

From that, by simple substitution,

$$\overline{W_e} = W + 2\bar{r}$$

Similarly, for equation (1.5.2) we get

$$\overline{W_e} = W + 2(\bar{r} - b)$$

Exercise 1.5.2. Fill in the following table as shown for the pyramidal cells:
  The general rule is

$$\text{pr\{pyramidal} \cap \text{stained\}} = \text{pr\{stained}|\text{ pyramidal\}} \cdot \text{pr\{pyramidal\}}$$

Table A.1.1

Cell type	Among 100 cells, there exist	In Golgi staining we expect to find	Percentage of cells seen in Golgi staining
Pyramidal	75	$75 \cdot 2\% = 1.5$	$100(1.5/2.09) = 72\%$
Smooth stellate	15		21%
Spiny stellate	8		6%
Mortinotti	2		1%
Total	100	2.09	100%

Exercise 1.5.3 You may try to work out Table A.1.1 in the reverse order, or you may use Bayes's formula:

$$\text{pr}\{\text{type } i\} = \frac{\text{pr}\{\text{type } i|\text{stained}\}}{\text{pr}\{\text{stained}|\text{type } i\} \cdot \sum_i (\text{pr}\{\text{type } i| \text{ stained}\}\text{pr}\{\text{stained}|\text{type } i\}}$$

You should get 60% pyramidal, 16% smooth stellates, 16% spiny stellates, and 8% Martinotti cells.

Exercise 2.3.1. We seek the smallest m for which

$$\sum_{i=0}^{m-1} \text{pr}\{i; 4.5\} \geq 0.94$$

Answer: $m = 9$.

Exercise 2.3.2. The expected number of synapses, S, as a function of the distance between the centers of the dendritic domain and axonal range can be evaluated by

$$E[S] = \int_x^x dx \int_0^x dy \, 2\pi(a_n a_m/\rho)y \exp[-k(R_1 + R_2)]$$

where x, y, R_1, and R_2 are found in Figure A.2.1. Numerical evaluation of this equation yields Figure A.2.2.

Exercise 2.3.3. (1) Compute the density of the axonal synapses.

$\nu = 12,000/0.098 \simeq 120,000/\text{m}^3$

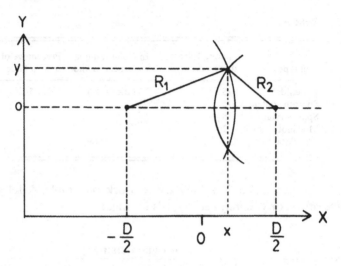

Figure A.2.1. Computing the number of synapses for an annulus with the radius y located at a distance x.

(2) Integrate the density of the axodendritic synapses over the cylinder (axonal range) V.

$$S = (4\pi v a_m/\rho) \int_{rcV} r^2 e^{-kr} dr \simeq 2.5$$

Note that

$$\int_0^{0.25} r^2 e^{-kr} dr < \int_{rcV} r^2 e^{-kr} dr < \int_0^{\infty} r^2 e^{-kr} dr$$

Exercise 2.4.1. (1) The density of the thalamo–spiny stellate (TSS) synapse is

$$\rho_{ij} = 4.8 \cdot 10^7/mm^3$$

The density of TSS synapses given by one axon is

$$v_{ij} = 6.7 \cdot 10^4/mm^3$$

The density of TSS synapses received by a spiny stellate cell is

$$\mu_{ij}(r) = 4.4 \cdot 10^5 e^{-r/0.06}/mm^3$$

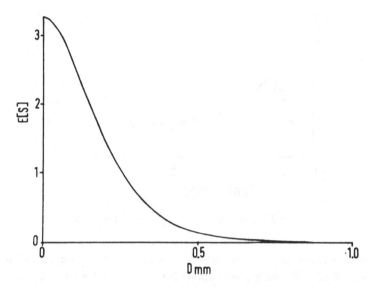

Figure A.2.2. Expected number of synapses as a function of the separation. For $E[S] = 0.06$ we get $D = 575$ μm.

(2) If we choose a spiny stellate cell at the center of the axonal range of a thalamic afferent, we expect to find $\simeq 2$ synapses between that particular axon and the stellate cell.

Exercise 3.2.1. Our null hypothesis is that there is no synapse between the cells. If we measure for T seconds, we expect to observe $5T$ spikes from one of the cells. After each of these spikes, we measure the activity of the other cell for 0.01 s. During this time we expect to see

$$x = 5T \cdot 0.01 \cdot 5 = 0.25T$$

spikes of the other cell by chance. The standard deviation for that number of counts is

$$\sigma = \sqrt{X} = 0.5\sqrt{T}$$

If there were a synapse (with ASG = 0.01), it would produce an excess of

$$\Delta = 5T \cdot \text{ASG} = 0.05T$$

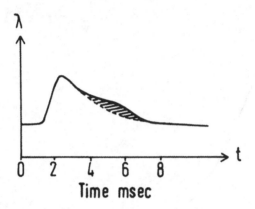

Figure A.3.1. Synaptic response curve after direct stimulation.

spikes in the second cell. We can reject the null hypothesis with confidence of 0.99 when Δ/σ is about 2.5 (as can be found in a table of the normal Gaussian distribution). That is,

$$2.5 = \Delta/\sigma = 0.05T/0.05\sqrt{T} = 0.1\sqrt{T}$$

from which we get

$$T = 625 \text{ s}$$

Exercise 3.2.2. We wish to find the ASG for which

$$(\lambda_1 \cdot \text{ASG} \cdot T_1)/(\lambda_1\lambda_2\Delta tT)^{1/2} \geq 2.5$$

which yields $\text{ASG} \geq 0.05$.

Exercise 3.2.3. (A) After direct stimulation of a_1, we expect to find the cross-correlation shown in Figure A.3.1. The second hump has, at most, an area of $(\text{ASG}_1)^2 \cdot \text{ASG}_2$, which is generally (for the cortex) negligibly small. Usually it will be even smaller than this maximum, because the somata of both a_1 and a_2 are being activated while in their refractory period. (B) When the ongoing activity is cross-correlated, we expect to see both synaptic response curves (Figure A.3.2). For positive delays, we see the synaptic response curve for a_1 to a_2. For negative delays, we see the mirror image of the synaptic response curve scaled by λ_1/λ_2.

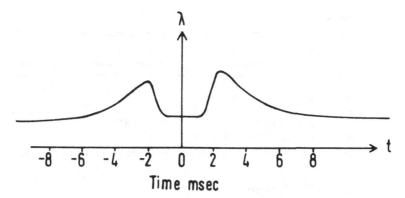

Figure A.3.2. Cross-correlation between a_1 and a_2.

Exercise 3.3.1. The gain through one chain is $(0.2)^{10} \simeq 10^{-7}$. We will need 10 million(!) such chains in parallel to see one extra spike at the end.

Exercise 3.3.2. Construct Table A.3.1. The required state is that at the third time step, the number of excitatory cells return to K. That will happen if

$$(0.8EN)^2 = 0.2EIN^2$$

from which

$$I/E = 0.64/0.2 \simeq 3$$

The inhibition must be threefold stronger than the excitation.

Exercise 3.3.3. x, the expected number of synchronous spikes, is 1.

$$\text{pr}\{k; 1\} = \frac{1}{ek!}$$

$$\text{ASG} = \frac{0.01}{e} \sum_{k=0}^{200} \frac{2^{k-1}}{k!} \simeq 0.014$$

Exercise 3.3.4. The gain when n interneurons are present is $n\text{ASG}^2$.

Table A.3.1

Time	Number of excitatory neurons	Number of inhibitor neurons	Comments
−1	K	L	Steady state
0	K	L	
1	$K + \Delta$	L	Perturbation
2	$K + 0.8\Delta EN$	$L + 0.2\Delta EN$	
3	$K + \Delta(0.8EN)^2 - \Delta0.2EIN^2$		

Exercise 3.3.5. Hint:

$$\sum_{i=0}^{\infty} i^2/i! = 2e$$

We get ASG = 0.02.

Exercise 3.4.1. Hint: Write equations for

pr{a spike of a_0 during $[\tau, \tau + d\tau)$}
pr{an extra spike of a_1 during $[t, t + dt)$ | given a spike of a_0 during $[\tau, \tau + d\tau)$}

Then introduce a large number of stimuli N and compute dn and dm.

Exercise 3.5.1. Hint: Show that

$$t^n e^{-kt} * e^{-kt} = \frac{t^{n+1}e^{-kt}}{n + 1}$$

Graphically the result looks like Figure A.3.3.

Exercise 3.5.2. Peak time = $10 \cdot$ (synaptic delay) + $9/k$. For a synaptic delay of 1 ms and decay time constant of 5 ms, the delay to the peak will be 55 ms!

Exercise 4.2.1. The average (variance) of the sum of n independent processes is the sum of the averages (variances). Therefore,

$$\mu = \sum_{i=1}^{n} \lambda_i A_i k_i$$

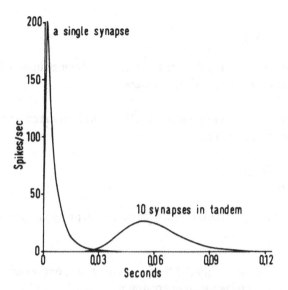

Figure A.3.3. One synapse and ten synapses in tandem. The ASG of the synapses is assumed to be 1. The decay constant of the synaptic effects is assumed to be 5 ms.

$$\sigma^2 = \sum_{i=1}^{n} \lambda_i A_i^2 k_i / 2$$

Exercise 4.2.2. If all the A_i and k_i have the same values A and k, respectively, then

$$\sigma^2 = \sum_{i=1}^{n} \lambda_i A_i^2 k_i / 2 = A^2(k/2) \sum_{i=1}^{n} \lambda_i = A^2(k/2)\overline{n\lambda}$$

where n is the number of fibers converging on the postsynaptic neuron, and λ is their average firing rate.

Answer: The conditions described in Exercise 4.2.2 do not affect the estimate of the amplitude of the postsynaptic potentials.

Exercise 4.2.3. Let us denote by $\overline{k}$ the average decay constant (assuming that all neurons fire at the same rate λ).

$$\overline{k} = \frac{1}{n} \sum_{i=1}^{m} k_i$$

Therefore,

$$\sigma^2 = n\lambda \overline{A^2 k}/2$$

If the inputs differ only by their time constants, then equation (4.2.5) can be used with the average time constant.

Exercise 4.2.4. Let us denote by $\overline{A^2}$ the mean squared amplitude of the postsynaptic potentials.

$$\overline{A^2} = \frac{1}{n}\sum_{i=1}^{n} A_i^2$$

Then, with all other parameters being equal for all presynaptic neurons,

$$\sigma^2 = n\lambda \overline{A^2} k/2$$

Note that $\overline{A^2}$ can never be smaller than $\bar{A}^2$. Therefore, our estimate of A in Example 4.2.1 is probably an overestimate.

Exercise 4.2.5. Periodic with the same period as the original waveform.

Exercise 4.2.6. A pulse of σ^2 at zero delay, and zero everywhere else.

Exercise 4.2.7. An equilateral triangle of height 1 and base $2T$.

Exercise 4.2.8. (1) Because of the factor $1/2T$ in equation (4.2.7), its autocorrelation is zero. (2) $\lambda r(\tau)$.

Exercise 5.4.1. If the inhibition is multiplicative, then (on a logarithmic scale) the amount of shift of the I curves is proportional to the firing rate of the inhibitory neurons.

Exercise 5.5.1. (1)

0	−1	−1
−1	0	1
−1	1	0

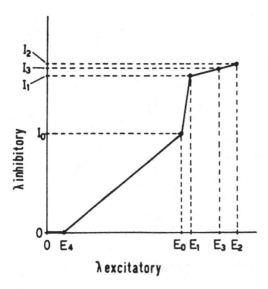

Figure A.5.1. Firing rate of the inhibitory neurons at step $n + 1$ as a function of the firing rate of the excitatory neurons at step n.

(2)

State	Discontent
$-1, -1, -1$	1
$-1, -1,\ 1$	2
$-1,\ 1, -1$	2
$-1,\ 1,\ 1$	-1
$1, -1, -1$	-5
$1, -1,\ 1$	0
$1,\ 1, -1$	0
$1,\ 1,\ 1$	1

$(1, -1, -1)$ and $(-1, 1, 1)$ are putative stable states. (3) See Figures A.5.2 and A.5.3.

Exercise 5.5.2. (1)

$$
\begin{array}{rrr}
0 & 0 & 1 \\
-1 & 0 & 0 \\
0 & -1 & 0
\end{array}
$$

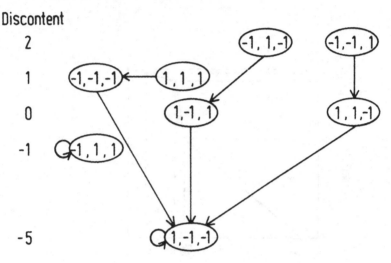

Figure A.5.2. Behavior when updating is parallel.

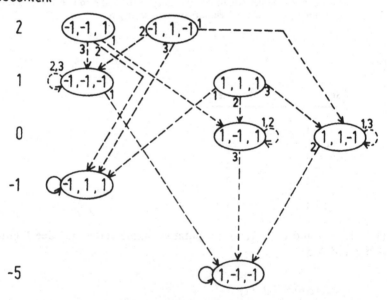

Figure A.5.3. Behavior when updating is random. The network's states are plotted according to their discontent. The graph illustrates the rule that for a symmetric J, a change in the firing state of one neuron must take the network's state downhill.

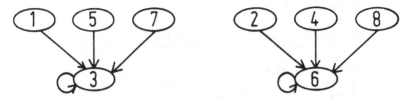

Figure A.5.4. Behavior of the network when updating is done serially in the order 1, 2, 3. States 3 and 6 are attractors.

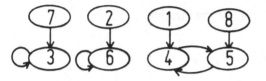

Figure A.5.5. Behavior of the network when updating is done serially in the order 1, 3, 2. Note that there exists a limit cycle even though the updating is done serially.

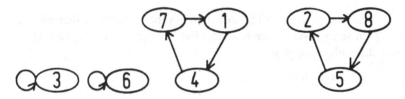

Figure A.5.6. Behavior of the network when updating is done in parallel.

(2)

Constellation	State number
(−1, −1, −1)	1
(−1, −1, 1)	2
(−1, 1, −1)	3
(−1, 1, 1)	4
(1, −1, −1)	5
(1, −1, 1)	6
(1, 1, −1)	7
(1, 1, 1)	8

(3) See Figures A.5.4, A.5.5, and A.5.6.

Exercise 5.5.3.

Network state	Number
(1, 1, 1, 1, −1, −1, −1, −1)	1
(1, 1, −1, −1, −1, −1, 1, 1)	2
(1, 1, −1, −1, 1, 1, −1, −1)	3
(1, −1, −1, 1, 1, −1, −1, 1)	4
(1, −1, 1, −1, 1, −1, 1, −1)	5

Note that these five states are strictly orthogonal.

Exercise 5.5.4. At the threshold,

$$\theta g_i = \sum_{j \in E} J_{ij} S_j - \theta \sum_{j \in I} J_{ij} S_j$$

which means that the difference between the depolarizing current ($\Sigma_E JS$) and the shunted current ($\theta \Sigma_I JS$) must be equal to the current needed to bring the resting membrane to threshold (θg_i).

Exercise 6.1.1. (1) The expected number of neurons that will die during ten years in one chain is almost five. The probability of finding an intact chain after ten years is

pr{0; 5} = 0.007

We look for w (number of parallel chains) such that

pr{0; 0.007w} < 0.01

Answer: $w > 600$.
(2) The expected survival is $100 \cdot (0.995)^{20} = 90.5$. Therefore, the expected number of deaths is 9.5 neurons per chain. The number of chains required is

$$W \geq -\ln 0.01/e^{-9.5} \simeq 60{,}000$$

Exercise 6.2.1. How about Figure A.6.1?

Exercise 6.2.2. See Figure A.6.2.

Exercise 6.2.3. See Figure A.6.3.

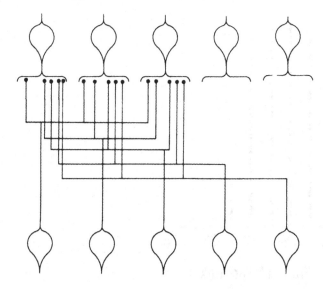

Figure A.6.1. Each neuron in the sending node establishes synaptic contacts with three neurons in the receiving node.

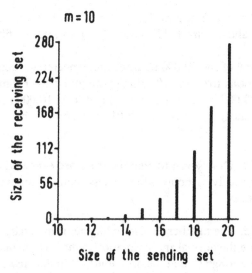

Figure A.6.2. Expected number of neurons that receive ten connections from a sending set of W cells.

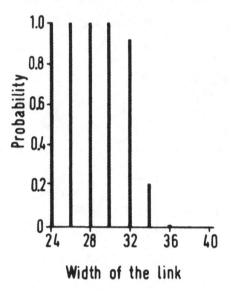

Figure A.6.3. Probability of finding a diverging/converging link having a multiplicity of 50 percent by chance.

Exercise 7.2.1. For a neuron firing at 5/s, the threshold is 2.33 standard deviations above threshold. Table A.7.1 should be filled in first.

Then find how many of the 20,000 neurons are expected to receive at least 1/SG50 connections from the 70 firing pyramidal cells. From that, we see that among 20,000 neurons there will be at least 10,000 neurons responding synchronously if the ASG is 0.01, and none if the ASG is 0.001.

Exercise 7.3.1. Only seven to eight neurons are expected to fire synchronously at the receiving node, which is less than the ten neurons that excited this node.

Exercise 7.3.2. The number of responding neurons is fairly large (50 or more), so we use the normal approximation to the Poisson distribution. We seek X large enough so that the probability of finding fewer than 50 neurons will be smaller than 0.01. This yields $X = 70$. In order to obtain an expected number of 70 neurons responding in the receiving node, we need to excite about 62 neurons in the sending node (Figure 7.3.1).

Answer: If we activate 62 or more neurons of the sending node, we are

Table A.7.1.

ASG	A	2.33/A	SG50	Number of neurons receiving at least 1/SG50 connections To 17 EPSPs
0.01				20,000
0.001				0

assured (to 99%) that the activity in the receiving node will not fall below 51.

Exercise 7.4.1. The probability of getting an EPSP at $[\tau, \tau + d\tau)$ is $f(\tau)d\tau$. The contribution of an EPSP at time τ to the depolarization measured at time t is EPSP $(t - \tau)$. The expected depolarization at time t is

$$\int_{-\infty}^{t} f(\tau)\text{EPSP}(t - \tau)\, d\tau$$

Exercise 7.4.2. Convolution of a boxcar having a width of 0.1 ms with a ramp having a duration of 1 ms will have a parabolic shape at the section $[0, 0.1]$, a linear rise in the section $[0.1, 0.9]$, and again the mirror image of the parabola in $[0, 0.1]$ at the section $[0.9, 1.0]$. With the values given in the exercise, the slope of the linear part is 1.

Exercise 7.5.1. The ignition threshold is 25; therefore, when 25 (out of 100) neurons fire synchronously at the sending node, we expect that 25 (out of 100) of the neurons in the receiving node will fire. Let A denote the amplitude of the EPSP at a neuron in the receiving node. Show that each neuron at the receiving node is depolarized (on the average) by $7.5A$, and show that its probability of firing is then 0.25. Show that when depolarized by this amount its membrane potential is thus 0.675 standard deviation below threshold.

At rest, its membrane potential is 2.33 standard deviations below threshold. Therefore, 7.5 EPSPs produce a depolarization of

$2.33 - 0.675 = 1.655$ standard deviations
$A = 1.655/7.5 = 0.22$ standard deviation
$$\text{SG50} = \frac{A}{T} = \frac{0.22}{2.33} = 0.095$$

Exercise 7.5.2. Denote the ignition threshold by T, and the probability of having fewer then T neurons firing simultaneously within 1 ms by P. The probability of failure within 40 ms should be less than 0.01. That is,

$$P^{40} \le 0.01$$
$$P \le 0.89$$

The number of neurons expected to fire each millisecond is $25 \cdot 200/1,000 = 5$. The threshold should be such that the probability of obtaining less than threshold (T), when the expected is five, should be less than 0.89.

$$\sum_{i=0}^{T-1} \mathrm{pr}\{i; 5\} \le 0.89$$

which yields $T = 8$.

Index

Printed in the United States
By Bookmasters